NASA

(National Aeronautics and Space Administration)

1958 onwards

COVER IMAGES: TOP: Mission Control, from where every US space mission since June 1965 has been managed. BOTTOM (left to right): the Vehicle Assembly Building at the Kennedy Space Center where Saturn V, versions of the Saturn IB and the Shuttle were prepared for flight and where the Space Launch System will be assembled; a Shuttle launch from LC-39; Flight crews ride the Shuttle into space on the Orbiter flight deck. *(NASA)*

First published in October 2018

A catalogue record for this book is available from the British Library.

ISBN 978 1 78521 115 7

Library of Congress control no. 2018935488

Published by Haynes Publishing,
Sparkford, Yeovil,
Somerset BA22 7JJ, UK.
Tel: 01963 440635
Int. tel: +44 1963 440635
Website: www.haynes.com

Haynes North America Inc.,
859 Lawrence Drive, Newbury Park,
California 91320, USA.

Printed in Malaysia.

Dedication

After the Apollo fire in 1967, NASA Flight Director Gene Kranz called a meeting of his staff in Mission Control and delivered an inspirational speech that was later reprised after the *Columbia* disaster in 2003:

From this day forward, Flight Control will be known by two words: Tough and Competent. Tough means we are forever accountable for what we do or what we fail to do. We will never again compromise our responsibilities. Competent means we will never take anything for granted. Mission Control will be perfect. When you leave this meeting today you will go to your office and the first thing you will do there is to write Tough and Competent on your blackboards. It will never be erased. Each day when you enter the room, these words will remind you of the price paid by Grissom, White, and Chaffee. These words are the price of admission to the ranks of Mission Control.

This book is dedicated to that spirit and to the men and women of NASA, who are tough, in their uncompromising approach, and competent, in not accepting failure, and daily do good work, inspiring new generations.

Acknowledgements

In preparing this book I would like to recognise an extremely large number of people at NASA who, over more than 50 years, have generously provided historical and reference material as a parallel to my professional activity at various field centres. Some of those people are no longer with us; the rest know who they are and I thank them.

In the production of this book, I would like to express appreciation to my commissioning editor Steve Rendle for his support, patience and guidance, to James Robertson for the design and layout, to Dean Rockett for proofing the text and to Peter Nicholson for the index. Anything good about the book is due to them. Any failings are down to me.

David Baker
East Sussex

NASA

(National Aeronautics and Space Administration)

1958 onwards

Operations Manual

An insight into NASA operations, programmes
and field-centre facilities

David Baker

Contents

OPPOSITE In America's most-visited thunderstorm State, NASA confounds nature and sets its sights on the Moon, the planets and eventually the stars. *(NASA)*

Introduction

Mention the word "NASA" and most people think of the Apollo Moon landings, while the majority of the rest will think of the Space Shuttle or perhaps the Hubble Space Telescope, Mars probes or the International Space Station. Watch TV? Hardly a day goes by without NASA being mentioned in the news or one of its spectacular missions explained in a dramatic documentary, while big-screen movie makers come to NASA for help in putting together blockbusters such as *The Martian*. But just what is NASA, when did it start, who pays for it and where is it located?

BELOW Arguably the most celebrated events in the history of the US space programme were the six Apollo Moon landings. Here, Alan Bean descends the ladder of the Lunar Module Intrepid during the Apollo 12 mission of November 1969 when he became the fourth man to set foot on the lunar surface. *(NASA-JSC)*

NASA – the National Aeronautics and Space Administration – is the agency of the United States government responsible for aeronautical research and for the development of space science, technology and engineering. It is also responsible for carrying out fundamental investigations into scientific questions posed by national institutions and scientific bodies for which research in the aeronautical or space sciences could provide answers.

In fulfilling its mandate, NASA is responsible for the development of satellites, space-based observatories and spacecraft and for the human exploration of the solar system. It is the prime US agency for the civilian exploration of space and for the application of satellites, space vehicles and technologies made available for the benefit of all.

It is the universal dissemination of information, freely and without prejudice, that underpins what NASA is and, in doing that, it represents the finest intentions of those politicians, and the executive branch of the US government, that set it up. To many people this is perhaps one of the most surprising aspects of NASA's mandate – that it is required by law to inform the general public about its work and to do that without compromising its integrity, neither lobbying for nor endorsing partisan causes or directly supporting any particular commercial or corporate objective.

Yet, it is funded by the American taxpayer to stimulate progress in the aeronautical and space sciences and to maintain national leadership in space research and exploration. Over time, NASA has achieved that through bi-partisan support in Congress and through the contribution it has made to the economy, to stimulating science, technology, engineering and mathematics in schools and to stimulating a rich breeding ground for undergraduates and graduates as well as providing a fruitful and rewarding career path for many people across the country.

Unanticipated at its inception, while fulfilling

these national objectives as a priority, NASA has provided leadership in international projects and programmes which have, in return, added to benefits for the United States of America, helping to bring organisations together who would be unable to accomplish great goals on their own. Now, through that leadership and with a sustained commitment to international cooperation, NASA is giving the USA the opportunity to engage with capabilities so great they are beyond the capacity of any one nation on Earth. Such capabilities are Moon bases and missions to Mars.

Beyond that, it is stimulating private and publicly owned commercial companies who seek to take advantage of the growing demand for services and supply chains, from launch vehicles to capsules carrying astronauts, and freighters carrying cargo to stations in space. And that too is a great return to the US economy, adding value to investment and raising financial benefits for all. In fact the overall financial value of a dollar spent on the space programme returns five-fold to the national economy – helping to pay for security, defence, health and welfare for the under-privileged.

NASA operates 13 field centres and facilities across the United States supporting a broad range of government-funded research and engineering programmes, adding science to exploration while stimulating invention, discovery and technology. A lot of which feeds out into the broader community of non-space-related industries, universities and academic institutes.

This is NASA – all you have marvelled at and a lot more besides.

ABOVE The Space Shuttle _Atlantis_ comes home. In 133 successful missions between 1981 and 2011, this remarkable flying machine enabled the assembly of the International Space Station and the launch of several satellites among which were great observatories and planetary missions. _(NASA-KSC)_

LEFT The exploration of Mars has been one of NASA's greatest success stories, represented here by the Mars Science Laboratory and the Curiosity roving vehicle, here engaged in traversing trials on undulating terrain. _(NASA-JPL)_

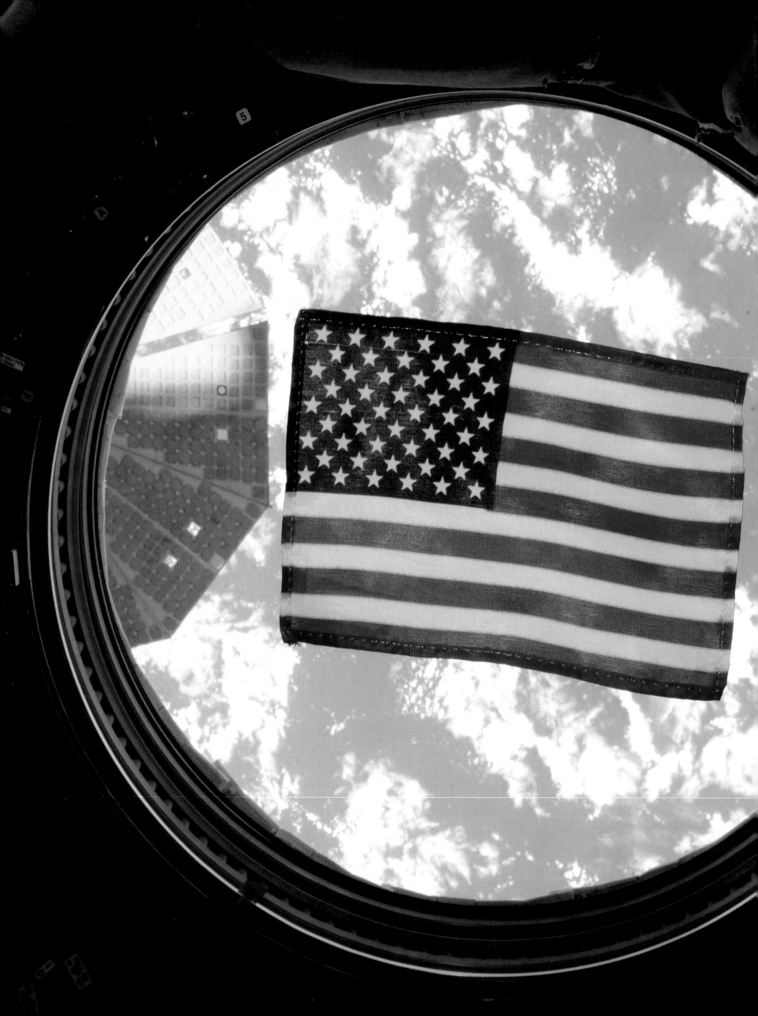

1 What is NASA?

NASA is a product of the early Space Age and was formally open for business from 1 October 1958. But its origins can be traced to 3 March 1915 when the Advisory Committee for Aeronautics (ACA) was formed. By this date, conflict in Europe that began on 4 August 1914 had spread fast to become the First World War, in which all forms of industrial machinery were employed in the great battles that quickly broke out. Among these machines were aeroplanes, flimsy devices made from wood and canvas powered by primitive forms of the internal combustion engine.

OPPOSITE The ultimate achievement. Since the beginning of this century there has never been a single moment when there has not been at least two humans in space, courtesy of NASA and the international Space Station. But there is more to the space agency than space itself, a journey that began more than 100 years ago. *(NASA)*

Almost from the outset, aeroplanes were employed for reconnaissance, gun-laying and surveillance of enemy territory, which made them targets for other aircraft aimed at denying the enemy the eyes which were vital for gaining an advantage on the battlefield. As other aeroplanes were employed on light bombing duties, the need to bring enemy aircraft down achieved an urgency unimaginable when conflict broke out.

The United States would not be involved in the war with Germany and the Central Powers until April 1917 but there was already concern in the US government that progress with aviation in European countries was outstripping that in America. Orville and Wilbur Wright, two bicycle-makers from Dayton, Ohio, had been the first to conduct a sustained flight with a powered aircraft on 17 December 1903. But, in the years that followed progress had been slow as the secretive Wright brothers, suspicious of their patented inventions being stolen, allowed European aviators to catch up.

With war accelerating the pace of development in Europe, the US was falling behind. In Britain, on 30 April 1909, the government had set up the Advisory Committee for Aeronautics and this was the model upon which the US Congress authorised funds for the NACA, the word "National" being inserted at its first meeting. Its mandate was to provide direction for the scientific study of flight and aeronautics and to determine which experimental techniques and devices were required to advance the state of the art as it applied to these new and urgent applications of aviation. Where once aeroplanes had been regarded as the playthings of the rich, they were now important instruments of war.

There had been false starts already. President Taft had been instrumental in proposing a National Aerodynamical Laboratory

Commission in December 1912, voted down the following month in Congress. Only when Charles D. Walcott, then secretary of the Smithsonian Institution, took up the challenge and recruited Senator Benjamin R. Tillman and Representative Ernest W. Roberts to carry the message in each House did success ensue. Paradoxically, it was the US Navy that saw the importance of the legislation, offering to attach it to the Naval Appropriations Bill which carried through the NACA legislation almost unnoticed!

Initially, the Committee had 12 unpaid members and received an annual budget of $5,000 in its first year, sufficient for one paid staff member. Not until 1917 did the NACA get its first facility – the Langley Memorial Aeronautical Laboratory in Hampton, Virginia – which opened for business in 1917 and from where it could begin to conduct original research. This was followed by the Ames Aeronautical Laboratory at Moffett Field, California, in 1939, the Aircraft Engine Research Laboratory at Brook Field, Ohio, in 1942 and the Muroc Flight Test Unit, California, in 1945. Over time the size of the NACA grew, with 100 employees in 1922 rising to more than 400 in 1938 and 650 in 1940. By the end of the Second World War it had 6,800 paid employees.

The NACA encouraged "blue-sky" thinking, which stimulated original research added to practical development of aerofoil shapes and

ABOVE Only slowly did the Army begin to take an interest in the possibilities opened up by aviation. Orville Wright demonstrates the A-model at Fort Meyer on 9 September 1908. Eight days later it crashed killing the passenger Lt Thomas Etholen Selfridge, the first person to die in an air accident. *(US Army)*

BELOW Publishing magnate William Randolph Hearst offered a prize of $50,000 to the first man to fly coast-to-coast across the United States, won by Calbraith Perry Rodgers between 17 September and 10 December 1911 on which date he taxied into the Pacific Ocean and settled any doubts about whether America was rapidly becoming an air-minded nation. *(David Baker)*

ABOVE A celebrated anthropologist and archaeologist, Secretary of the Smithsonian Institution from 1907, Charles Doolittle Walcott (1850–1927) was renowned for his discovery of Cambrian fossils in the Burgess shale, one of the most famous finds of all time. Seen here with his family, he convened a conference in 1914 supporting the authorisation of a government research facility which was established in 1915 as the National Advisory Committee for Aeronautics. *(NASA)*

ABOVE President William Howard Taft tried in 1912 to create a government research facility for the scientific investigation of flight but was voted down by Congress. *(Library of Congress)*

a wide range of inventions and discoveries from wind tunnel tests and flight experiments. Throughout the 1920s and 1930s, the US aviation industry benefitted greatly from the work carried out at these centres and the science of flight was advanced in no small measure by the

technical developments in industry underpinned by research at the NACA. It became a byword for the testing, verification and qualification of new discoveries in aerodynamics and in the scientific principles of flight.

During and after the Second World War, the NACA made numerous contributions to the expansion of commercial aviation and to national defence through a sustained effort

RIGHT The first meeting of the National Advisory Committee for Aeronautics (NACA) which formally came into being on 3 March 1915, almost exactly seven months after the outbreak of the war in Europe, rapidly fanned into the Great War, known thereafter as the First World War. *(NACA)*

ABOVE The seal of the NACA showing the first flight of a powered aeroplane controlled by a pilot which was on 17 December 1903 with Orville Wright at the controls. *(NASA)*

ABOVE The only aircraft available to the United States Army were a collection of Wright and Curtiss examples, typical of which was the Wright B, a replica of which is in the National Air and Space Museum in Washington, DC. *(David Baker)*

to break down barriers and achieve higher speed, greater altitude, improved efficiency and enhanced safety. Toward the end of the war, the NACA made a major contribution to the US Army Air Force's desire to find a solution to transonic compressibility and to demonstrate that supersonic flight was both feasible and practical for a new generation of combat aircraft powered by the jet engine.

To demonstrate the feasibility of supersonic flight, the Bell XS-1 employed a multi-chamber rocket engine to break through the mistakenly named "sound barrier" and to achieve the first manned flight in excess of Mach 1 on 14 October 1947. Less than one month after the formation of the independent US Air Force on 18 September, this flight proved that a new era had dawned in which it was entirely feasible to anticipate combat aircraft flying several times faster than anything before and to achieve altitudes above that which the propeller-driven warplanes of World War Two could achieve. None of that would have been possible without the support from the NACA for Air Force and Navy aircraft.

And so began a shared era between the NACA and the military in development in high performance aircraft that continues to this day, albeit with industry now taking over much of

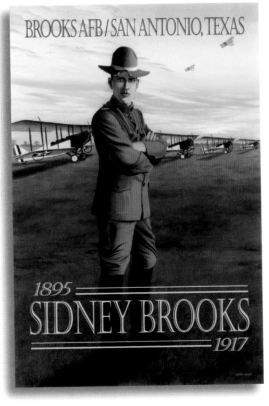

LEFT The logo of the National Advisory Committee for Aeronautics which was proudly worn as evidence of the first serious attempt by the US government to accelerate research into aviation and catch up with the Europeans. *(NASA)*

LEFT Sidney Brooks was an early casualty of military flying when he died in his Curtiss JN-4 on 13 November 1917, seven months after the United States entered the war in Europe. He would give his name to Brooks Air Force Base in San Antonio, Texas. *(USAF)*

LEFT The US Army had to buy almost all its aircraft from the French and the British, types such as this S.E.5a being manufactured at Wolseley Motors in Birmingham on a hand-built basis reminiscent of quality motor cars of the period. *(Bruce Robertson)*

BELOW Ordered into production to supplement the Curtiss JN-4 types, the Standard J-series trainers helped American pilots get their wings to fly foreign aircraft. *(USAF)*

RIGHT The Executive Committee of the NACA meets with American aircraft manufacturers in 1917 to put together a research and development programme for supporting US industry in establishing a competent aeronautical capability. *(NACA)*

ABOVE While industry lagged behind, held back by a lack of early preparation, America's pilots were carving a name for themselves in European skies as men like the famous Eddie Rickenbacker became air aces virtually overnight. *(USAF)*

ABOVE The NACA quickly picked up the pace for conducting basic research into aerodynamics and optimum wing aerofoil sections for aircraft such as this Boeing P-12 fighter developed in the late 1920s. *(USAF)*

ABOVE A contemporary of the P-12, the Curtiss P-6 Hawk relied on basic design options allowing manufacturers to benefit from research applicable to all industry contenders for government aircraft orders. *(USAF)*

the development work supporting new and innovative technologies and their application to aviation. The sustained research programme in supersonic flight that began with the XS-1 was continued with additional research types of which arguably the most famous is the North American Aviation X-15. Conceived in the mid-1950s, the hypersonic X-15 achieved speeds greater than Mach 5 and an altitude in excess of 300,000ft (91.44km).

In 1945, research on rockets was facilitated through the Pilotless Aircraft Research Station at Wallops Island, Virginia, under the aegis of the Langley Laboratory, followed two years later by the High Speed Flight Station located at the southern edge of Edwards Air Force Base in southern California. Rockets had come to play an impending role of great magnitude. Shaped by the Second World War, rocketry had focused on solid propellant devices for helping aircraft get in the air and for propelling munitions. But in late 1945 it was quickly seen as a means by which

RIGHT First of the monoplane fighters, the powerful little Boeing P-26 served with seven front-line pursuit groups. *(USAF)*

RIGHT The peak of advanced aircraft engineering was the emerging generation of heavy bombers which would underpin long-range bombardment, the Douglas XB-19 appearing just before the outbreak of another conflict in Europe at the end of the 1930s. (USAF)

BELOW The NACA helped industry explore unique and innovative design concepts, typified here by this Beech C-43 Traveler, in which the upper wing was placed aft of the lower wing, a type utilised by the Air Force as a light transport aircraft. (USAF)

the characteristics of the upper atmosphere and near-Earth space could be explored.

As German V-2 rockets began to arrive in the United States the Army and the Navy took a special interest in these ballistic devices, seeing in them the prototypes of a new generation of weaponry which could be used on the battlefield or to bombard land installations from ships. Studies supported by the NACA, however, saw in rockets the ability to carry instruments to altitudes greater than that achieved by balloons and to conduct detailed scientific surveys of the upper atmosphere which would, indirectly, support the development of new generations of aircraft for both civil and military applications. The rocket could, believed the NACA, help define

requirements for a new generation of high-flying aircraft for both civil and military purposes.

If the X-15 could be considered to be its "toe in the water" at the edge of the vast ocean of space, the next several years would see that involvement grow until, by 1957, approximately 40–50% of its work was generally space related. While the NACA was assisting the Air Force with development of this rocket-powered research aircraft, a decision had been made to launch artificial satellites to orbit the Earth as part of the International Geophysical Year (IGY) which began on 1 July 1957 and was intended to last 12 months until extended to 31 December 1958. It followed two International Polar years held in 1882–83 and 1932–33.

In the United States there had been a determination to maintain a peaceful intention to the outside world and for that reason the Vanguard satellite project which was set up to support the IGY was managed by the Naval Research Laboratory. As far as the general public were aware, this highly publicised effort was the only space project underway in the United States but in reality that was far from true. The decision to give the NRL management of Vanguard was made in August 1955, by which time President Dwight D. Eisenhower had already authorised the development of a clandestine spy satellite programme given the code name Corona.

Eventually, when launches began Corona would be revealed as the Discoverer programme, proclaimed as a scientific investigation of outer space by the military for a better understanding of the space environment when, in reality, it was a system for obtaining images of secret installations in the Soviet Union. The Vanguard project supporting the IGY was a useful diplomatic pathfinder to the spy flights which would follow. Eisenhower was concerned about Soviet reaction to objects flying over Russian territory. They had already shot down several aircraft that strayed into Soviet airspace and Eisenhower was worried that a satellite flying overhead may be interpreted as an act of war.

However, the spy satellite flights were several years away from launch and Vanguard was expected to be the world's first satellite to orbit the Earth. The NACA made a significant contribution to Vanguard and had in fact provided much detailed research data on the recovery of ballistic nose cones. Supporting the peaceful study of the Earth and its environment within the solar system, the first stage would be provided by the Martin Company which had produced the Viking sounding rocket – a "sounding rocket" being a ballistic rocket making soundings of the upper atmosphere through packages of instruments parachuted back to Earth, much like an instrumented line thrown over the side of a ship making soundings of the ocean.

In fact, satellites were considered the next step beyond sounding rockets – instrumented packages which would continue to orbit the Earth for very long periods without further need of propulsion, sending back information modulated on to the carrier wave of a radio signal. There was very great interest in this at

LEFT Dr Robert Goddard opened a new chapter in American flight history with the first launch of a liquid propellant rocket on 16 March 1926, another US "first" which would, again, not receive government support for at least a decade. *(NASM)*

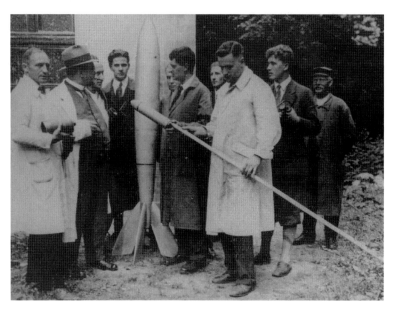

CENTRE Goddard with (to his immediate left) Charles Lindbergh, the first man to cross the Atlantic Ocean solo. Goddard would be ridiculed by some and encouraged by others but his reluctance to share research results frustrated many. *(Via David Baker)*

RIGHT In other places, other enthusiasts were experimenting with rocketry. Second from right, a bushy-haired Wernher von Braun became an eager member of the German *Verein für Raumschiffarhrt*, the German Society for Space Travel. *(Via David Baker)*

RIGHT Throughout the 1930s, the NACA provided a nationwide support service for aeronautics, engineering and research that greatly benefitted US aircraft manufacturers and associated industries in everything from better lubricants to standard aerofoil sections numbered by an NACA prefix. *(NACA)*

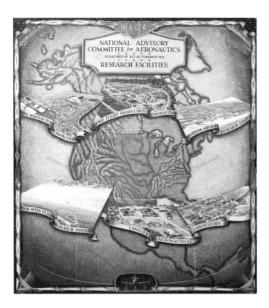

ABOVE General Henry H. "Hap" Arnold (1886–1950) played a pivotal role in boosting US development of jet-engined aircraft when he negotiated possession of all the research work undertaken in the UK by Frank Whittle and his Power Jets company which had resulted in the flight of a jet aircraft on 15 May 1941. Arnold saw to it that the NACA and the Air Force got their hands on one of the most advanced technologies of its time. *(USAF)*

LEFT Working with design details and blueprints from the UK, the Air Force worked with Bell to produce the jet-powered Bell P-59 Airacomet, here seen in the wind tunnel at the Langley Aeronautical Laboratory. *(NACA)*

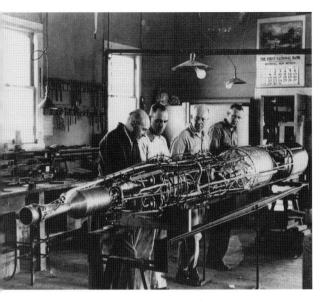

ABOVE Throughout the late 1930s, Robert Goddard had been working to develop increasingly more capable rocket vehicles which could be used to send instrumented science packages to the upper atmosphere, perfecting the design, increasing performance and also attracting a little interest from the NACA. *(Via David Baker)*

ABOVE RIGHT Had Goddard been channelled into perfecting long-range rockets, instead of being pulled away by the government into developing rocket booster packages for helping heavy aircraft get airborne, the entire history of the next 20 years could have been different as America sprinted to catch up with Russia. *(Via David Baker)*

OPPOSITE The P-59 first flew on 1 October 1942 but although only 66 were built it was a superb attempt to produce a practical combat aircraft, owing much to work on the project from NACA facilities. *(USAF)*

RIGHT A subscale test model of the German V-2 rocket, the A-5 represented the structural layout of the ballistic missile which would bombard London and Antwerp in the closing months of the Second World War. *(Via David Baker)*

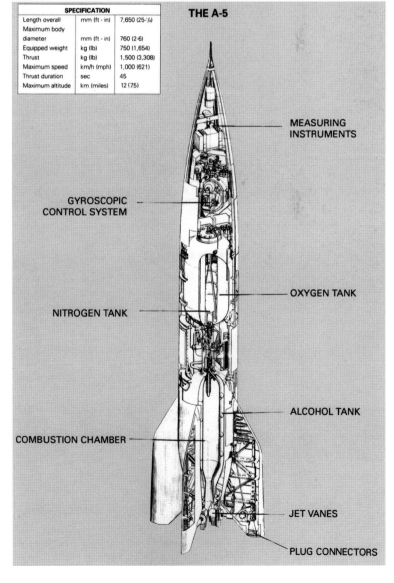

THE A-5

SPECIFICATION		
Length overall	mm (ft - in)	7,650 (25-¼)
Maximum body diameter	mm (ft - in)	760 (2-6)
Equipped weight	kg (lb)	750 (1,654)
Thrust	kg (lb)	1,500 (3,308)
Maximum speed	km/h (mph)	1,000 (621)
Thrust duration	sec	45
Maximum altitude	km (miles)	12 (75)

MEASURING INSTRUMENTS

GYROSCOPIC CONTROL SYSTEM

OXYGEN TANK

NITROGEN TANK

ALCOHOL TANK

COMBUSTION CHAMBER

JET VANES

PLUG CONNECTORS

LEFT A grainy photograph showing three road-mobile V-2 rockets ready for firing, its transportability making it a difficult target for pre-launch attack by air. *(Via David Baker)*

BELOW Project Paperclip allowed an initial tranche of 104 German rocket and jet propulsion scientists to be moved to Fort Bliss, Texas, after the war, sharing their technical skills with the victors who would soon build on these capabilities. *(NASA)*

BELOW During the Second World War, the NACA moved to play an active role in the development of research aircraft specifically designed to carry out research and gather data on high-speed and high-altitude environments, the Air Force's Bell XS-1 being one of the most prominent. First through Mach 1 in level flight, the Bell XS-1 "Glamorous Glennis", recognising test pilot "Chuck" Yeager's wife, displays its bullet-like forward fuselage section. *(NACA)*

BELOW Just noticeable on the fin of the second Bell XS-1 is the NACA logo applied to this aircraft which was used exclusively to conduct research flights studying the transonic and supersonic regimes. *(NACA)*

the NACA and research flights from the Pilotless Aircraft Research Station bore testimony to that. This added work had supported some growth within the NACA so that by 1955 it had 7,600 employees and a budget of almost $57million ($513million in 2018 money).

The shock of Sputnik

It is highly unlikely that NASA would exist had it not been for the launch of Sputnik 1 on 4 October 1957. Shocked by the surprise launch of the world's first artificial satellite by the Soviet Union, there was an immediate clamour for a strong reaction to what many Americans felt was an impending threat every bit as real as the surprise Japanese attack on the naval base at Pearl Harbor had been on 7 December 1941.

This fear was based on the fact that in 1949 the Russians had detonated their first atomic weapon much earlier than expected,

following that with a thermonuclear bomb in 1955. Now, Sputnik 1 demonstrated that they had the rocket with the power to deliver a weapon of unimaginable destruction to targets anywhere across the United States; truthfully, they reasoned that a rocket that could place a satellite in orbit could be used as an intercontinental ballistic missile (ICBM).

First launched on 15 May 1957, the R-7 was indeed Russia's first ICBM, several months ahead of America's equivalent, the Atlas missile. It was the R-7 which had been used to place Sputnik 1 in orbit. America's Vanguard was a very much smaller rocket, emphasising only too clearly the advanced lead held by the Soviets. These comparisons were an important barometer of public opinion in 1957 and technological advantage was a yardstick by which to measure the potential to win a war with the Soviet Union.

This was not an unreasonable assumption, especially at the peak of Cold War tensions.

ABOVE A proliferation of engines and rocket motors, including a reconstructed V-2 rocket which became the hallmark of the expanding spheres of interest for the NACA in the decade after the Second World War. *(NASA)*

ABOVE The first reassembled V-2 rocket erected at White Sands Missile Test Range, New Mexico, on 10 March 1946 in preparation for an engine test. *(US Army)*

Only the year before, the Russians had sent tanks into Hungary to quell an uprising which sought independent elections against the totalitarian communist government, its secret police in hock to Moscow. As 400,000 Hungarians fled to the West, about 30,000 were killed when Russian tanks rolled through the streets of Budapest. Many Americans

felt Russia to be an existential threat to world peace and the launch of Sputnik 1 only added to that fear. Moreover in a survey carried out by the US Information Service, only in the UK did people across Western Europe – then a divided continent – believe that the United States remained superior to the Soviet Union in technological prowess.

ABOVE RIGHT A V-2 rocket sits in a checkout gantry prior to flights which frequently carried instrumentation provided by the NACA for upper atmosphere research. *(NASA)*

RIGHT Active use of facilities at Edwards Air Force Base provided an expanding dataset into flight in the transonic and supersonic regimes. Here a Douglas Skyrocket, the first aircraft to exceed Mach 2, is posed with carrier-aircraft and chase planes. *(NASA)*

Thus it was that this single scientific step, relatively harmless in reality, defined events which would see the IGY, instigated to bring nations together not long after the most devastating conflict in history, transformed into a race to demonstrate superiority in advanced engineering and technology. From the moment that Sputnik began to orbit the Earth there were only two players in the race and the consequences of losing would send political and propaganda shock waves around the world. Prestige and the expression of power had been the only reason that Soviet premier Nikita Khrushchev approved the request from Sergei Korolev to use the R-7 he had designed to launch a satellite. This sent an unequivocal message to the United States and the rest of the world that they were not the race of uneducated and illiterate peasants from the Urals that most Americans thought them to be.

Yet, paradoxically, the Americans were already some way ahead of the Russians in plans for utilising access to space for military purposes; in addition to the CIA's Corona spy-satellite project, both the newly independent Air Force and the Army had ideas about how to exploit the space environment and were developing vehicles to achieve dominance in this new domain. And they were none too secret about it either, publicly boasting of ambitious concepts for using space to achieve military domination over rogue states and missile threats, not all of which were practical.

The Air Force had its Dyna-Soar (an abbreviation of "dynamic soaring") boost-glider designed to launch off an adapted ICBM such as Titan and conduct reconnaissance and strike missions anywhere on Earth within an hour using the lift capabilities of the winged vehicle to extend the trajectory. Dyna-Soar was evolving from a plethora of studies, each focusing on a specific military mission in space. Beginning in 1956 the Air Force had started on its Man-In-Space-Soonest (MISS) project designed to use a conventional rocket to lift a pilot on a ballistic trajectory to study the physiological effects of space flight imposed by Dyna-Soar. Following that, orbital flights would study the effects of prolonged weightlessness on the body.

The Army had the German V-2 rocket engineer Wernher von Braun, brought to

the United States at the end of the Second World War and already responsible for several battlefield and theatre missiles. In April 1957 his team at the Army Ballistic Missile Agency (ABMA) began studying rocket designs capable of lifting massive payloads of up to 39,690lb (18,000kg) into orbit, a requirement defined by the Department of Defense, far in excess of the 3,087lb (1,400kg) promised by rockets such as Atlas and Titan ICBMs then in development but not yet flown.

Developed exclusively as a satellite launcher for the IGY, America's Vanguard rocket was capable of putting a 20lb (9kg) satellite in orbit while Russia's Sputnik 1 weighed 184lb (84kg) and the disparity was clear, emphasised when Russia launched Sputnik 2, weighing 1,121lb (508kg), on 3 November 1957. The bleeping sphere that was Sputnik 1 had been merely a demonstration shot; the second carried a dog called Laika into orbit, a flight from which it would not return as the satellite had no means of getting back through the atmosphere without burning up.

The consequences of all this weighed heavily on the Eisenhower administration and all eyes were on Vanguard when a test firing on 6 December 1957 lasted only two seconds before the rocket and its 3lb (1.36kg) satellite fell back on to pad LC-18A at Cape Canaveral, erupting in a ball of fire. The team protested that this was only a test shot but the rocket did carry a

ABOVE In the mid-1950s the NACA provided support for the Vanguard satellite project, the US contribution to the International Geophysical Year of 1957–58. Managed largely by the Naval Research Laboratory seen here, which had been steered into space research by Ernst Krause, it would play a significant role in pushing America into the Space Age when most of the Vanguard personnel went to the Goddard Space Flight Center. *(NRL)*

satellite and the world knew the implication of its failure. The media had a field day, naming it "flopnik" to Russia's Sputnik! But failure here gave von Braun and his Army team of rocket engineers the chance to succeed where the "civilian" Vanguard project had failed.

An enthusiastic space advocate since boyhood, von Braun had claimed in 1956 that the Redstone ballistic missile he had developed for the Army could be modified to launch a satellite into orbit ahead of the Vanguard programme, which had run into several development problems. This had been denied so as not to send the message that the US military was behind the satellite effort. However, when Vanguard failed on 6 December, the Army was given the go-ahead and on 31 January 1958 Explorer 1 was successfully launched, sending America's first satellite into orbit. In fairness, Vanguard came good and was eventually successful in sending three satellites into space on separate occasions in 1958 and 1959.

The immediacy of the shock of Sputniks 1 and 2, however, galvanised a reaction

RIGHT While publicity surrounded futuristic projects which would never fly, the real surge in US space activity was within a range of highly classified military satellites, the most famous of which was the Corona spy satellite. Launched initially by a Thor Agena rocket, it consisted of a photographic system recording images on film returned to Earth in a recoverable capsule. *(USAF)*

BELOW While the US was moving slowly toward the anticipated launch of a Vanguard satellite, Russia put the first artificial satellite, Sputnik 1, in orbit on 4 October 1957 with an R-7 rocket shown here. *(Via David Baker)*

BELOW The reaction to Sputnik was electric, galvanising public concern about the potential military threat of surprise attack to a nation which only 16 years previously had suffered at the hands of a shock strike on Pearl Harbor. *(Via David Baker)*

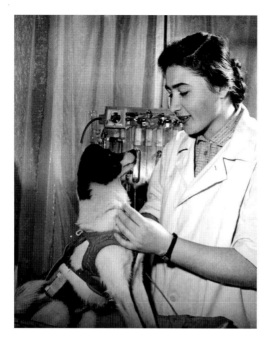

RIGHT Before the US could respond, the dog Laika was sent into orbit in Sputnik 2 on 3 November 1957. With a weight of 1,121lb (518kg), it demonstrated that the R-7 rocket was capable of delivering a nuclear warhead on the United States. *(Via David Baker)*

expressed in very different ways: while the public were either greatly concerned or largely indifferent, the military saw it as a reason for them to seize control of existing plans – those outlined earlier in the chapter – and to organise even more ambitious projects to go head-to-head with the Russians in an all-out bid for

BELOW Laika could not be recovered, as the Russians had no means of returning the capsule to Earth, and fear adopted hatred as animal rights groups around the world turned Soviet publicity into a negative stunt. Both Russian and American scientists used animals in high altitude test flights but there had always been a means of getting them to safety from suborbital ballistic flights. *(Novosti)*

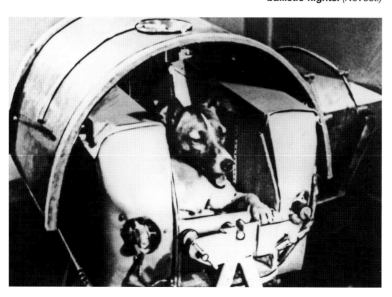

supremacy in this new arena. The politicians saw it as an affront to American global pre-eminence, self-evidently achieved with the end of the Second World War and the dropping of the atomic bomb, and sought ways to orchestrate a positive response engaging the scientists and engineers of America in an organised effort to put America back on top.

The only organisation central to these disciplines was the NACA, comprised of a Main Committee of 17 members: five from the Department of Defense, five from other government departments and seven from non-government bodies. Five technical committees and their 23 subcommittees boasted 450 members from government agencies, private corporations and key academic institutions. They had no part in the decision structure which was exclusively the preserve of the Main Committee. Regulations allowed for a chairman and a vice-chairman, with the chairman appointing what were referred to as the Top Three: a Director, an Executive Secretary, and an Associate Director for Research.

Far from inferring a triumvirate, the Top Three did not have a split leadership, although a division of direction was experienced when Dr Hugh L Dryden joined as Director from the Bureau of Standards in 1947 and found discord with Executive Secretary John Victory, who had been with the NACA since 1915. Eventually, and only with the help of some senior officials in the organisation, Dryden was able to exert his leadership and to be recognised for the role he played. Dryden was a generalist, with specialisation the forte of the Executive Director and the Associate Director for Research.

While the NACA had a very high proportion of its work dedicated to space research, engineering and technology insofar as these supported national programmes conducted by other more appropriate departments of the government, there was constructive opposition to its tilt away from pure aeronautics. Many NACA engineers and scientists had been drawn to work on the missile programmes, solving problems regarding propulsion, aero-thermal challenges of re-entry and control of vehicles at the edge of space such as the X-15 and the Vanguard programme. Now, the Sputnik challenge swung many more toward the

challenges of space exploration, their position strengthened by events.

Convinced that these rapidly evolving challenges would engage the NACA in a big way, Dryden and Chairman Jimmy Doolittle organised a dinner at the Hotel Statler for 18 December 1957. Several "third echelon" employees were invited to attend as the potential leadership which would have to construct a national space programme for the United States. After reviewing various courses of action, there was overwhelming support for Dryden's view that space would become a dominant function for the NACA.

ABOVE Called the Stever Committee after its chairman, Guyford Stever, NACA's Special Committee on Space Technology meets at Lewis, 26 May, 1958. Left to right, Edward R. Sharp, Director of Lewis Laboratory; Colonel Norman C. Appold, USAF; Abraham Hyatt; Hendrik W. Bode, Bell Telephone Laboratories; W. Randolph Lovelace II, Lovelace Foundation for Medical Education and Research; S. K. Hoffman, Rocketdyne Division, North American Aviation; Milton U. Clauser, Director, Aeronautical Research Laboratory, The Ramo-Wooldridge Corporation; H. Julian Allen, NACA Ames; Robert R. Gilruth, NACA Langley; J. R. Dempsey, Convair-Astronautics; Carl B. Palmer, Secretary to Committee, NACA Headquarters; H. Guyford Stever; Hugh L. Dryden (ex officio), Director, NACA; Dale R. Corson, Cornell University; Abe Silverstein, NACA Lewis; Wernher von Braun, Director, Development Operations Division, Army Ballistic Missile Agency. *(NASA)*

Inaugurating NASA

The very fact that there was no national space programme, only a disparate set of plans and possibilities proposed and campaigned for by the Army, the Navy, the Air Force, the NACA and a range of other competing interests, allowed a clean sheet upon which to write a structured plan for the future. Nobody was more aware of that than Eisenhower, who had made a speech to the nation two days after Sputnik 2 in which he announced that the country had made a major "breakthrough" in that it had recently successfully recovered a nose cone from a ballistic missile, solving the problem of getting an ICBM warhead back down through the

atmosphere to its assigned target. Not much to brag about – but it was something.

But bragging rights there were aplenty for the various armed services boasting of their "experience" in this field, each vying for the lead role as a national space agency. Sputnik had prompted Eisenhower to upgrade the Scientific Advisory Committee (SAC), formed in 1951 under the tenure of President Truman, and to attach the prefix "President", thus the PSAC became the most important line of communication between all these competing bodies and the White House. President of the Massachusetts Institute of Technology, James R. Killian left that post to head up the revitalised committee which became a highly significant voice for science at the top of the executive branch of government.

This summary of how the NACA addressed the challenges from Sputnik is far too brief to recognise all the meetings and debates which occupied anyone and everyone with even the vaguest interest in space over the following weeks. Developments divided into two specific categories: examination of existing activities and discussion of the nation's long-range plan for space activity. Because of the highly classified nature of the most robust, existing, programme (Corona) only a very select few were aware of the full breadth of just what was already underway. But the specifics of individual programmes mattered less than the synthesis of the debate which quickly focused on just two issues: the degree to which future space activities should be run by a civilian organisation, or by the military.

On 21 November the Rocket and Satellite Research Panel of the National Academy of Sciences under its Chairman Dr James A. Van Allen met and supported a National Space Establishment and on 4 December it was made known that the American Rocket Society had proposed such an organisation on 14 October and these two groups combined to push for that in a joint proposal on 4 January 1958. Exactly ten days later, the NACA penned a report suggesting an interagency body comprising the Department of Defense (DoD), the NACA, the National Science Foundation and the National Academy of Sciences. And then it was the turn of the politicians.

Throughout November and December 1957 and January 1958, hearings before the Military Preparedness Subcommittee of the Senate Committee on Armed Services struggled to define the difference between ballistic missiles and Earth-orbiting satellites, seeing the latter as very different to the former. This helped prepare the way for the Eisenhower administration to adopt the same approach to the challenge of Sputnik as it had to the call for an IGY satellite – to keep it within the civilian sector so as not to "militarise" the next frontier before it had been settled. But it had not been his first inclination to do that. Only when Vice President Richard Nixon and James R. Killian argued against that did Eisenhower acquiesce and agree to it being an open, civilian agency.

Arguably the most important meeting of all in a continuous succession of such events was held on 25 November 1957 when Senator Lyndon B Johnson, Chairman of the Senate Committee on Armed Services, opened a series of hearings over 20 days at which more than 70 witnesses, most from the DoD, produced more than 2,300 pages of testimony on what to do next. There were so many different points of view that a flood of bills were put before Congress, including one from Senator Clinton Anderson that sought to give the Atomic Energy Commission the major share in a future space programme.

In rejecting that and the other bills, the White House felt incumbent to come up with its own proposition and on 4 February 1958 President Eisenhower announced that he had charged James Killian and the PSAC with finding a solution. Based on conclusions reached, on 5 March 1958 the President approved the recommendations of his Advisory Committee on Government Organization that leadership in space research should rest with the NACA, endorsing a direction in which Killian, the PSAC and the Bureau of the Budget had been moving for some time.

On 16 January 1958 the Main Committee of the NACA passed a resolution recommending a joint programme such as that previously espoused by Dr Dryden and recommending an increase in staff from 8,000 to 17,000 over a three-year period with a corresponding increase in its budget from $80million to $180million a year. When it appeared, the March recommendations carried little of the original NACA concept for a future structural organisation. Liabilities listed as to the NACA's lack of competence in across-the-board space activity in that most of US space development was already under way in the military, the NACA was not in a position to move ahead with demonstration projects, it was limited by pay restrictions through civil service regulations, and its structure and organisation were geared to a much lower level of expenditure than that deemed necessary for a national space effort.

The 5 March memorandum recommended that the NACA be reorganised as the National Aeronautical and Space Agency, that it should be allowed to establish pay rates higher than the conventional restrictions on government pay, and that its head should be appointed by the US President. It stressed the urgency in Congress of drafting legislation, the need to work out relationships with the Department of Defense, the requirement for some existing activities to be transferred to the new agency and that it should emphatically be empowered with "the full responsibility for developing and arranging for the execution of the civil space program". The draft legislation was sent up to Congress on 2 April.

Congress had already set about organising itself for the coming wave of legislative activity. The Senate had created the Special Committee on Space and Astronautics on 6 February 1958, chaired by Senate majority leader Lyndon Johnson, while the House of Representatives set up the Select Committee on Astronautics and Space Exploration effective from 5 March under House majority leader John W. McCormack. One member was a future US President: Gerald R. Ford. The first of 17 days of House hearings involving statements from 51 individuals began on 15 April, resulting in testimony running to 1,541 pages in the public record.

When the draft legislation was sent to Congress on 2 April, President Eisenhower took a close personal interest in requiring the NACA and the DoD to "jointly review the pertinent (space) programs currently under way within or planned by the (DoD, to recommend)…which of these programs should be placed under the direction of the new Agency". Talks to achieve that were held by Dryden and the Deputy Secretary of Defense Donald A. Quarles.

ABOVE All change but business as usual as the NACA's High Speed Flight Research Station loses its logo at the transfer to NASA effective 1 October 1958. *(NASA)*

Immediate agreement was reached on transferring the space programmes of the Advanced Research Projects Agency (ARPA), set up on 7 February 1958 as another response to Sputnik. Established at the express order of the President, it was charged with ensuring no further "Pearl Harbor" moments for the United States by integrating research activity between universities, government departments and industry to harness new technologies and ensure that they were merged with practical applications. It was also understood that directly space-related defence facilities would be transferred to the new agency.

At this time those ARPA projects included the Vanguard satellite programme and some lunar probes then in the planning stage but differences of opinion surrounded the Air Force MISS project, the plan to send pilots into space in a capsule launched on top of an adapted missile such as Redstone or Atlas. Initially, the NACA and the DoD agreed to joint management of MISS but the Bureau of the Budget objected on the grounds that it would cost too much to have this duality and it was agreed to transfer that too. NASA would rename it Mercury, America's first manned space flight programme.

Setting up shop

Fearful of an escalating militarism, President Eisenhower had got his way – the national space initiative would be run by a civilian organisation built on the back of one of the world's most respected aeronautical research and advisory bodies, renowned for its cutting-edge contribution to advancing the science and engineering of flight and flying.

But things would be very different when the new body metamorphosed out of the NACA. Whereas the existing organisation had never really started and managed a major project, the new one would be expected to establish a wide range of objectives and see them through to completion, setting goals in Earth science, space science, lunar and planetary exploration and in human space flight, which for a while would soar in importance and come to represent the very embodiment of exploration, carrying the torch for American supremacy on the high frontier of advanced science and engineering.

Because the organisational structure of the new NASA would be very different the leadership would have to change. For several months after the draft bill was sent to Congress on 2 April it was assumed that Dryden would remain its head. But in testimony to the new House space committee he had shown reluctance to sign up to a crash programme which might involve large sums of money on uncertain goals which might never be achieved. This was not what Congress wanted to hear. It wanted action – now and immediately. Moreover, Eisenhower was in the habit of putting Republicans in to major government positions and Dryden was a Democrat.

It fell to James Killian to find a new boss for the emergent NASA. On 7 August 1958 he asked T. Keith Glennan, the former president of Cleveland's Case Institute under whose leadership it had matured into one of America's top engineering schools, to come to Washington. Recipient of five honorary doctorates, Glennan was to face challenging times, often frustrated by the lack of a bequeathed space policy, like Dryden intent on not promising things he could not deliver but an untested administrator nevertheless. It was Eisenhower who gifted the job to Glennan, a Republican but selected on merit and not on his politics.

The Space Act was signed on 29 July. The successor to the NACA would be the National Aeronautics and Space Administration, not an Agency as originally planned, but the lowest level of government categorisation to which the head was a White House appointee. And there was a specific reason for that. NASA was to be an agency that would define policy and prepare a long-term plan. It was not the highest level, a "Department" (such as the Department of Defense, the Department of State or the Department of the Interior), nor was it at the second tier level, an "Agency" (such as the Central Intelligence Agency), but an Administration (a tier 3 level such as, later, the National Oceanic and Atmospheric Administration).

NASA's boss was the Administrator and the first person to hold that position was the one sent to the Senate on 9 August for confirmation, approved six days later, with Glennan as Administrator and Dryden as Deputy Administrator. They were sworn in at the White House on 19 August. A considerable amount of work remained and a notional date for transfer from the NACA to NASA was set for no earlier than 1 October 1958, sensibly so because that date represented the start of the second quarter of fiscal year 1959 (a financial year which in those days began on 1 July) and would make accounting and legislation neat and tidy.

But there was a lot to be done, as NASA prepared to receive $100million in space projects from DoD to NASA – not of course including the top secret Corona or related programmes then under the remit of the CIA and other organisations – and to lay

BELOW Seated at his desk at the Marshall Space Flight Center, Director Wernher von Braun directs a generation of Saturn rockets for which the military had no use but which will underpin a decade and more of outstanding achievements including the first landing on the Moon. *(NASA)*

out objectives for a national civilian space programme. Yet for all the paperwork and signed documents, the transfers would take several months with much more to implement in practice than had been signed over already. But it was a historic shift and employees who left work with the NACA on Tuesday 30 September came back the following day to report in as NASA employees.

Almost immediately, before long-range plans could be formulated, apart from Vanguard, the most advanced activity inherited from the Air Force was the MISS programme, promising human space flight to a nation hungry for achievement and ambitious goals. Of course MISS involved ballistic capsules carrying human guinea pigs for studying the effects of space flight on the body prior to operational use of the winged Dyna-Soar boost-glider for sustained and protracted missions. Ballistic capsules were not the path along which the NACA had believed the long-term development of manned space flight would evolve but, paradoxically, it was a method which would dominate the way astronauts went into space long after Dyna-Soar was cancelled in 1961 until the Shuttle launches began 23 years later.

Other transfers included two Army lunar probes and three satellite projects and on 14 October Glennan made a move to transfer those responsible facilities as well. The Army had consolidated its missile programmes into the Army Ordnance Missile Command (AOMC) with headquarters at Redstone Arsenal near Huntsville, Alabama. AOMC embraced three commands: the Army Ballistic Missile Agency (ABMA), the Army Rocket and Guided Missile Range (ARGMA) and the White Sands Missile Range (WSMR) at White Sands, New Mexico. Von Braun was Director of Technical Development of the Development Operations Division at the ABMA and as such had responsibility for some military missile programmes as well as the Jupiter and Saturn launch vehicles.

The Army also owned the Jet Propulsion Laboratory (JPL), which was staffed and managed by the California Institute of Technology (CalTech) and which reported to the AOMC. Together they had provided the Explorer 1, America's first orbiting satellite,

and the transfer to NASA of JPL and half the ABMA's Development Operations Division was requested. While the DoD supported the request the Army fiercely opposed such a move and lobbied strongly among its friends in Congress to retain what it believed to be in the national security interest of the country. While JPL was transferred to NASA at the end of 1958, the Army eventually agreed to allow NASA access to ABMA and AOMC units for its civilian space activities but things would change.

On New Year's Day 1959, NASA had 8,420 people, having acquired 150 from the Vanguard satellite programme, 50 from the Naval Research Laboratory and 200 from other locations but not until mid-1960 would it acquire its next large injection of personnel. By the end of January 1959 the general organization of NASA had been laid out. Arranging the NASA budget was complex because it was initiated a quarter of the way through fiscal year 1959 and at about the time discussions were under way with the Budget Bureau for FY1960. The President's January 1959 request to Congress included $485million for NASA. More on the budget and how that changed over time can be found in Chapter 4 of this book.

Throughout 1959 NASA formulated its long-range plan embracing lunar and planetary missions and manned flights around the Moon. It supported DoD development of a succession

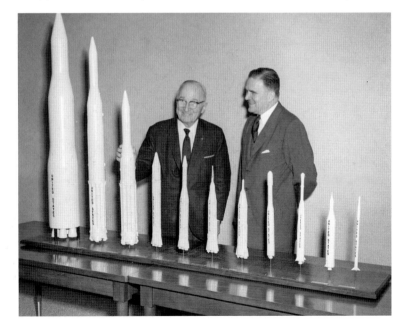

of new variants of existing rockets into satellite launchers, designed and developed an all-solid Scout satellite launcher, conducted conceptual study of a very large super-booster called Nova and took over management of a very powerful rocket motor, the F-1, under development by Rocketdyne and transferred to NASA from the Air Force. During that year NASA launched eight satellites and two Moon probes out of which a total of three were successful.

NASA's flagship was the one-man Mercury spacecraft, the renamed MISS project taken over from the Air Force, with a contractor selected to build it and seven astronauts selected for flights. It was also given a DX priority rating, the highest for any government project securing first call on resources, men and materials. A global tracking network was contracted to Western Electric in July and the agency began development of a deep-space communications capability for lunar and planetary missions. Less publicised, NASA acquired from the DoD the Tiros weather satellite programme.

Expansion

ABOVE Taking charge of programmes and projects previously started and managed by the Army and the Air Force, NASA supports its first initial forays into space, in this case marking the flight of Pioneer IV with von Braun (left) and H. Julian Van Allen (right), the scientist from Iowa State University who had provided the instruments for America's first satellite, Explorer 1, launched on 31 January 1958. *(NASA)*

During 1959 the steady growth in projects, programmes and personnel expanded the agency and defined its central role in space exploration. Staff levels increased to 9,567 by the end of the year and NASA had formulated a plan for the future exploration and application of space. Arguably one of the most important decisions made in 1959 was the construction of a new field installation dedicated to space research and the development of scientific satellites.

This had been planned long before the decision to transform the NACA into NASA and was implemented as a matter of urgency not long after the new agency began operations. Named the Goddard Space Flight Center, it was NASA's first dedicated facility built exclusively for space research and applications. But acquisitions too were to follow in 1960, adding greatly to NASA's operational capabilities.

The decision to retain the Army Ballistic Missile Agency in Army hands, which had been forced on NASA in December 1958, presaged several months of cooperation and Glennan

RIGHT Pioneer IV was launched by this Juno II rocket, another product of the former Redstone Arsenal, on 3 March 1959 to a trajectory that carried it to within 36,650 miles (58,983km) of the Moon the following day on its way to an orbit of the Sun. *(NASA)*

spoke highly of the Army's attitude to the civilian agency. But Congressional reluctance to allow NASA to prise apart space-related functions at the ABMA evaporated and during the summer of 1959 the DoD prepared a major reorganisation which was revealed on 23 September that year.

Under this revision of structure, the Army had no use for the massive Saturn rocket being developed by von Braun's team. The story of the Saturn I is covered in another Haynes book specifically detailing how that rocket evolved but it is important here to relate the outline of that story, especially as it would bequeath to NASA one of the jewels in its crown. Details can be found in the relevant facilities section later in this book but suffice here to say that in mid-1960 the von Braun team and its Saturn rocket programme was transferred to NASA and the facility renamed the George C. Marshall Space Flight Center.

A year later NASA opened the Michoud Test Facility (MTF) and in 1962 it began construction of a dedicated launch site on Merritt Island, Cape Canaveral, Florida, to handle flights of

the mighty Saturn V in support of the Apollo Moon programme, a facility later renamed the Kennedy Space Center. Marshall, the MTF and Kennedy facilities were part of the same evolution toward the design, manufacture and operation of large rockets and the ambitious objectives associated with them.

All this was something very new to an organisation that only four years earlier had been a small body of engineers and scientists working through three laboratories to advance the science of flight and flying, now finding themselves thrust into the foreground of exciting, adventurous and dramatic endeavours, most of which had been defined in the first year of operation. There had been no blueprint for the exploration of space but, far exceeding the remit of his job specification, Wernher von Braun had already been elevated to high status through extensive lobbying for space travel and the exploration of the Moon and Mars.

In books and articles for prestigious news-stand products such as *Colliers Magazine*, as consultant to film magnates such as Walt Disney and as adviser to the military and to government

ABOVE A priority for NASA was to pick up the US Air Force MISS programme and transform it into the Mercury manned flight effort which required new and adapted rockets to get its astronauts into orbit. This included development of a Little Joe spacecraft test rocket (right), an adapted version of the Redstone rocket for suborbital test shots (centre) and an Atlas launch vehicle for orbital flight. *(NASA)*

departments, the rise and rise of Wernher von Braun had been a sensation. Planning flights to the Moon and Mars since an adolescent youth in the early 1930s, he laid down a sequence of evolving capabilities which would carry humans deep into the solar system. First, he said, would come a space station assembled in Earth orbit, followed by flights to the Moon, and then lunar bases as precursor steps to the colonisation of Mars.

This "von Braun paradigm", as it has been called, became the unofficial blueprint sought by the insatiable appetite of an excited public and by the desires of politicians to make America first and best in the race for space. It all seemed very logical, a set of sequential steps, each one building on the capabilities of the former to construct a plan for expanding the human presence into space. It followed the rationale for rocket testing at the Redstone Arsenal where multi-stage launchers were to be tested with first the initial stage, then with two stages and finally with three stages – or however many were required in the "stack" – the informal term for the complete assembly of stages and payload on the launch pad.

The first long-range plans formulated by NASA followed this paradigm, with smaller, supporting projects and programmes along the way. But in the late 1950s, the presence of humans in space was considered essential; machines working autonomously on pre-

programmed instructions were not thought to be possible for all but the most basic of missions and self-determination or artificial intelligence was firmly in the realm of science fiction. To do anything big in space needed people, scientists, engineers, explorers, pilots, technicians and doctors. Progress with humans in space was considered to be an integral milestone on the path to exploration of the solar system beyond the simplest unmanned probes. This is why astronauts were so lauded from the time they were selected and why the race to put the first man in orbit was considered so strong a measure of progress in space; forget the satellites and not-so-successful Moon probes – it was people that wrote headlines in the world's newspapers. And that is why, when the Russian cosmonaut Yuri Gagarin beat the Americans into space it made such an impact around the globe. None more so than in the White House, soon smarting under a failed attempt to invade Cuba and overthrow the communist regime of Fidel Castro.

It was the combined blow of Gagarin's flight on 12 April 1961 and the failed Cuban invasion at the infamous Bay of Pigs starting five days later that caused the White House to react by seeking ways to upstage the Russians and propel the search for technological supremacy into an all-out space race. Inaugurated as President on 20 January 1961, John F. Kennedy had sought a new head for NASA while showing little interest in the space programme apart from granting a modest budget increase for the Saturn I rocket, still several months away from its first flight. James E. Webb became NASA's second Administrator after accepting an invitation from the President on 14 February.

Over the next several weeks President Kennedy asked Vice President Lyndon Johnson, largely instrumental in getting legislation hurried through for the National Aeronautics and Space Act and the establishment of NASA, to determine the best way to beat the Russians, using a series of ever more ambitious goals and selecting the one which was achievable by the United States but just beyond the reach, on a similar timescale, of the Russians. Options covered several bases from orbiting space stations to a landing on Mars but a Moon landing seemed just feasible and was assessed as being outside the capabilities of the Russians in the medium term.

NASA itself was divided on the issue, as only vague outlines of potential Moon landings had emerged from the long-range plans of 1959 and 1960 and they envisaged that bold step no sooner than the early 1970s. After intensive deliberations from the Atomic Energy Commission, the Budget Bureau and senior Congressional leaders, the recommendation was to go for a manned landing on the Moon by the end of the decade. Some wanted it accelerated to "before 1967", the 50th anniversary of the revolution that toppled the Russian autocracy in the belief that the Soviets would target their own landing to celebrate that event. Given the magnitude of the challenge, prudently that earlier date was not in the final call to action made by President Kennedy when he addressed a joint session of Congress on 25 May 1961 and announced the goal.

This was arguably the greatest challenge faced by NASA to this day. In 1961 there were no certainties about how to achieve that and they had little more than eight years to get astronauts on the Moon. Nobody knew for sure how to accomplish the task and none of the rockets required to achieve that goal had even been designed. Only the Russians had even put a man in Earth orbit and there were no space vehicles anywhere near being capable of sending humans beyond Earth and there were certainly no serious designs for putting men safely down on the lunar surface.

Paradoxically, the only organisation that had given serious technical study to the subject was the British Interplanetary Society (BIS), formed in 1933 and boasting some of the most brilliant brains in rocketry among its membership. Before the Second World War in 1939 and immediately after the end of that conflict in 1945, scientists and engineers at the BIS had drawn up detailed plans which envisaged a four-legged lander capable of putting men on the surface. They had prepared engineering drawings which through logical conclusion presaged the design eventually selected for NASA's lunar lander. As usual, in solving engineering challenges and finding solutions to ambitious goals, the BIS was way ahead of its time.

The decision to go for the Moon left

undecided the manner in which the flight would be conducted and three solutions were proposed: a Direct Ascent (DA) using a massive Nova super-booster capable of lifting directly off the surface of the Earth and heading for the Moon, separating from the terminal stage of the rocket and setting down on the surface before lifting off and returning to Earth; several smaller Saturn-class rockets assembling in Earth orbit a single vehicle which could then carry out the steps for the Direct Ascent mode, a plan known as Earth Orbit Rendezvous (EOR); or the Lunar Orbit Rendezvous mode (LOR).

The idea of LOR was a relative newcomer, comparatively unknown and challenged by many but discussed in some circles and brought to the fore through the untiring and unremitting persistence of John C. Houbolt, an engineer from the Langley Research Center.

BELOW An early "first" for NASA in the field of planetary exploration was the flight of Mariner 2 which, on 14 December 1962, conducted a close fly-by of Venus and returned information about its environment which astounded scientists and began to transform an awareness of its intensely dense and hot atmosphere, leading to a conclusion enshrined in the "greenhouse effect" applied later to the Earth. *(NASA)*

It required a second spacecraft to carry the astronauts down to the surface of the Moon while the main spacecraft remained in lunar orbit awaiting the return of the crew. Finally, 14 months after the decision to go to the Moon had been made, NASA signed off on the LOR mode in July 1962.

LOR required two spacecraft, not one, to fulfil the mandate and in November 1962 Grumman was selected to build what was initially called the Lunar Excursion Module (LEM), "Excursion" being dropped within two years. In looking to the future, NASA had wanted to develop a three-man spacecraft which it had named Apollo in 1960 but the Eisenhower administration had refused to allow NASA, then only two years old, to develop a spacecraft to succeed the one-man Mercury capsule until early missions indicated just how useful man would be over machine.

Ever concerned to balance the books, Eisenhower had resisted the push to make the space programme a permanent and expensive part of government, preferring to take simple, logical steps rather than rush to grand objectives. His successor was not so restrained but when the Moon goal was announced NASA built its plan around existing studies on the three-man Apollo, took those designs off the shelf and awarded a development contract to North American Rockwell (NAR), the company that had built the X-15, on 28 November 1961.

For the next seven months NASA and NAR believed Apollo would be the spacecraft to put astronauts on the Moon but the LOR mode changed all that.

However, realising that whatever mode was eventually selected would need an interim spacecraft to develop the operational techniques to be used on the Moon missions, on 7 December 1961 NASA issued a contract to McDonnell Aircraft for development of their one-man Mercury into a two-man spacecraft called Gemini. This would prove to be one of the most rapid programmes ever mounted by NASA with development, manufacture and all 12 flights completed within five years.

Mercury was good for little more than a day in orbit and a Moon mission would last up to two weeks so testing men and machines for long-duration flight was urgent; rendezvous and docking would be essential and that needed to be tested in space by bringing two separately launched vehicles together and linking them up; space suits would be needed for astronauts to go outside their spacecraft; and special life-support backpacks would be essential for keeping the spacewalking crewmen alive. All these would be successfully tested by ten manned flights in the Gemini programme carried out during 1965 and 1966.

NASA redefined

Precipitated by the Moon goal and accompanied by a massive escalation in NASA funding (see Chapter 4), all these expanded programmes and operations required a significant change in the managerial structure of NASA and this was instituted at the end of 1961. Nobody could have foreseen the expansion of NASA and the colossal challenges it faced. Almost immediately after the announcement by President Kennedy it was apparent that a new facility, a Manned Spacecraft Center, was needed to replace Mission Control located for Project Mercury at Cape Canaveral Air Force Station, Florida. It would provide a home for the Space Task Group from the Langley Research Center which had masterminded America's first human space flight programme. That home would be located near Houston, Texas, and achieve fame around

BELOW In response to the flight of Russia's Yuri Gagarin, President Kennedy challenged NASA to land men on the Moon by the end of the 1960s, resulting in a search for suitable landing craft among which was this concept leading to the Apollo Lunar Module. *(NASA)*

the world as the site of NASA's human space flight operations.

The changes introduced on 1 September were the first of many structural shifts to adapt to changing requirements over the next several decades but this one set the standard for continuous evolution. Administrator Webb worked the change to give NASA clearer focus and to provide greater emphasis on the agency's major programmes. It would also provide the centre directors "an increased voice in policy making and program decisions". To achieve that, all centres would be placed under the direction of the Associate Administrator who was then Robert C. Seamans, assisted by a Deputy Associate Administrator. To this date NASA had four programme offices: Advanced Research Programs (OARP); Space Flight Programs (OSFP); Launch Vehicle Programs (OLVP); and Life Science Programs (OLSP).

From 1 November these four programme offices were abolished and in their place the following headquarters offices were established: Advanced Research & Technology (OART under Ira A. Abbott); Space Sciences (OSS under Homer E. Newell); Manned Space Flight (OMSF under D. Brainerd Holmes); and Applications (OA with the vacancy as yet unfilled). In addition, there was a supporting office for Tracking and Data Acquisition (OTDA under Edmond C. Buckley). The importance of getting the NASA message across was now to be under an Assistant Administrator for Public Affairs reporting directly to Webb.

These changes began to move NASA away from the inherited management team from NACA and moved people in from industry; for instance, Holmes came from RCA. And there was deliberate intention there. Others selected for higher things declined because they felt they could not carry out higher office under the new structure, Abe Silverstein being one who accepted a job as director of the Lewis facility rather than take on the top job at the Office of Manned Space Flight.

Administrative changes are never as attention-grabbing as rocket launches or Moon landings but they help explain why certain situations emerge as a result of gradual evolution from specific decisions about

authority and control. Previous to the November 1961 changes, each field centre reported to headquarters, as they always had done under the structure of the NACA. Now, they would be focused in direction as to what programmes they were to manage but achieve much greater freedom in that they received a greater level of autonomy and would report higher up the chain of command at headquarters but they still looked to general management at NASA for their resources.

On top of all these organisational changes, NASA effectively set about the task of reporting on its activities and informing the general public on progress within selective programmes. Anyone familiar with NASA's public information people today would never understand the culture prevalent in the 1960s and 1970s. Most public affairs staff were from the military, men who had covered conflict and been a part of combat units during the Second World War or in the Korean War of 1950–53. They understood the dynamic of action and the need to tell people the truth, unsullied by "corporate-speak" or hidden agendas. They were articulate in winning the confidence of the electronic and print media who they marshalled and mobilised, always searching for truth and reality.

ABOVE To prepare for Moon missions, NASA developed the two-man Gemini vehicle based on the earlier one-man Mercury spacecraft. Beginning flights in March 1965, on the second flight Ed White became the first American to walk in space on 3 June 1965. *(NASA)*

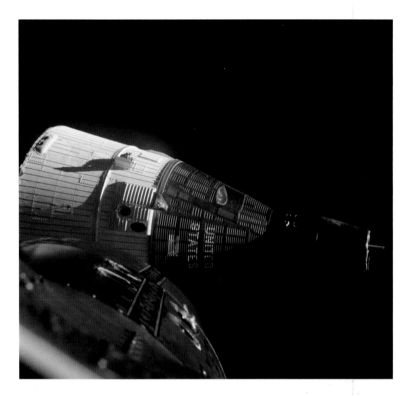

ABOVE A secondary requirement for the Gemini programme was to demonstrate that astronauts could remain in space for up to 14 days, sufficient for an Apollo moon mission. This was accomplished in December 1965 by Gemini 7, visited in orbit by Gemini 6 as seen here. *(NASA)*

BELOW Rendezvous and docking was essential if the Lunar Module was to re-dock with the Apollo mothership after a lunar landing so Gemini 8 demonstrated that a docking between two spacecraft was feasible. *(NASA)*

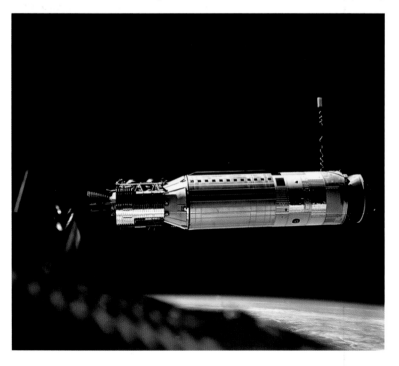

But there was a legal requirement for NASA to fully explain its actions to Congress, the legislature that controlled final decisions over budgets, and to keep the White House, the executive which would implement those acts, fully in the loop, supporting the requirement for open disclosure. Regular press conferences were held to inform the public about progress with specific programmes, projects or budgets and gradually the barrier that had held the media at a distance during the early days at Cape Canaveral, for instance where the emphasis had been on missiles and cruise weapons testing, evaporated and even the Air Force came to see the value that would accrue from maintaining a close and trusting relationship with the press, opening their doors wherever possible.

Film reports, usually 15 minutes in length, were issued at quarterly intervals by each field centre and through specific programme offices summarising progress with particular programmes. These became useful for TV stations to show them to their audiences as a way of displaying access and satisfying the increasing demand for greater insight to these exciting projects. These filmed quarterly reports are now available on YouTube and for download from numerous websites and are an unequalled insight to the detailed and in-depth descriptions of technical research and development back in the 1960s and 1970s. The level of sophistication was far higher than it is today. Where once they were informative and technical, PR agencies now pump up excitement by broadcasting style over substance.

One significant outcome from the late 1961 restructuring of NASA was the universal adoption of PERT – Program Evaluation and Review Technique – which replaced the PDP (Program Management System) which had been in place from the old NACA days. One of the great developments in project planning, oversight and control had been the shifts which had been essential to managing giant projects such as the Manhattan programme to develop America's atomic bomb. As well as being more sophisticated, PERT was capable of being computerised and could accommodate a very large number of variables while setting down milestones for charting progress. It was capable of handling a complex set of events in multiple chains and of charting a critical path through specifically identified nodes.

But PERT was a milestone checker and early-warning net for project managers to use in identifying upcoming roadblocks in the critical path or providing alternate routes avoiding blocks which could stall progress. It did nothing to evaluate the quality of the management decision or to help those plans be set down more effectively but it was the best tool available for helping improve the overall management of a system or of a programme.

PERT came to NASA via Walter Haase, a US Navy specialist, through a meeting he had with Seamans on 17 January 1961 in which he outlined the advantages to the agency of adopting such a system. It had been developed for the Navy's Polaris ballistic missile programme, which had demonstrated unequalled speed in bringing an idea to operational readiness as America's sea-based leg of the nuclear deterrent triad. The full Navy version had been tailored to that programme and what Haas recommended was an adapted version, which was fully adopted by NASA on 1 September 1961. It was difficult to induct at NASA and eventually, in 1962, Management Systems Corporation was brought in to implement a training programme.

If all this sounds a little too academic to have relevance to the main thrust of the NASA story, suffice to say that the June 1961 Fleming Report on options for a manned lunar landing as prepared fully utilising PERT and that this had been instrumental in drawing up a realistic timetable for implementing the landing before the end of the decade. However, when Holmes was appointed boss of NASA's Office of Manned Space Flight in that November he avoided PERT, disliking it to a level where he baulked at its extolled virtues and failed to make proper use of it. That, in turn, precipitated the second major organisational change at NASA, one which would turn around the flagging Apollo programme and would eventually, through decisions made in 1963, see Americans on the Moon as intended by President Kennedy.

Critical to achieving the manned lunar landing by the end of the decade was an explosive escalation in the NASA budget. While that is

more fully described in Chapter 4, it is important to acknowledge here that the management of a budget far greater than anything envisaged by the NACA was as crucial an element in getting men on the Moon as any of the more exciting technical developments. But that was only the tip of an increasingly complicated pyramid involving an organisational structure which would expand NASA personnel to more than 30,000 people coordinating the products of several tens of thousands of contractors and subcontractors who directly employed more than 450,000 engineers and scientists on NASA projects, most of it on manned flight goals.

For a thin sliver of time, between November 1962 (when Grumman got the contract to build the LM) and May 1963 (the end of the Mercury programme) NASA was responsible for four manned space flight vehicles: Mercury, Gemini, Apollo and the Lunar Module. Thereafter, until the end of the Gemini programme in December 1966, it had three simultaneously in work. That was a major challenge to the organisational infrastructure but that in turn would bring challenges when the sheer size of NASA contracted in the aftermath of Apollo. But the manned Moon landing was only one programme among many that NASA ran during these heady days and that, in itself, brought

ABOVE During the heady days of the Gemini programme, NASA successfully flew the first mission to reach Mars, flying past the planet and sending back 21 pictures in 1965 from Mariner 4. *(NASA)*

RIGHT President Lyndon B. Johnson, who had masterminded legislation to found NASA out of the NACA and who had convinced President Kennedy to send men to the Moon, receives a portfolio of Mariner 4 pictures from Director of the Jet Propulsion Laboratory William Pickering. (NASA)

a difference of opinion between the senior leadership and the White House.

President Kennedy had never been an ardent enthusiast for a diverse space programme and his only action on entering the White House had been a small budget boost to accelerate development of the Saturn I launch vehicle. Since Sputnik 1, the measure of pace in the space race had been calibrated by the potency of rockets and their lift capacity and Saturn I would achieve that parity with Russia. As though this yardstick alone had been for Kennedy the only point in funding NASA, after the first flight of Saturn I on 27 October 1961 he began to question his own decision in laying down the Moon challenge.

Soundings were made through the United Nations, where Kennedy threw down the option of "going to the Moon together" with Russia, and in confidential discussions where senior NASA leaders were called in to discuss such a possibility. Without any real technical understanding of rockets and space vehicles, for Kennedy the space programme was a political tool to pick up and put down as necessary and in that he was not so very different at all from every other President. But Kennedy lacked the foresight to connect with any space project other than Apollo and in discussions with Jim Webb in the Oval Office, he chastised the NASA boss for attempting

to build a broad space programme. For Kennedy, no other activity at NASA should take precedence over this Moon goal and he demanded that of Webb.

To some extent this was unavoidable and while total NASA expenditure soared so too did the percentage allocated to manned space flight, increasing from 51% in 1962 to 61% in 1963 and 66% in 1964. Many science programmes begun prior to the May 1961 Moon decision were absorbed and diverted to serve this single goal, especially unmanned lunar exploration where so little was known about conditions at the surface. Ironically, unknown to anyone outside the Soviet Union, Russia would not approve a Moon landing programme until 1964, the year after the assassination of John F. Kennedy.

Meanwhile, the decision over NASA's management of the Apollo programme came into doubt, at first from middle management and then at the very top. The organisation changes effected in late 1961 were causing problems and the overall pace which had been anticipated was not happening. There was a serious danger of not achieving the Moon landing goal solely through reasons of conflicting management concepts. Changes began to be made in 1962, when Kennedy was first having serious misgivings and Congress was being faced with a much higher NASA budget request, ironically from the White House, than anticipated. It was time to replace the "old" NASA with a more effective evolution.

NASA evolves

On the promise that he could completely restructure NASA, George E. Mueller agreed to leave a well-paid job in industry and join NASA as head of manned space flight, replacing Holmes who left under a cloud, some of which was largely due to challenges beyond the capacity of one man to solve. Mueller took over on 1 September 1963 and what ensued was a significant shift in NASA fortunes.

Mueller introduced his "GEM boxes", named after his initials, and set out to significantly improve relations between the field centres, which were becoming fiefdoms in their own right rather than operating under the disciplined constraints of headquarters. Famous for having an uncompromising attitude toward his managers by organising regular Sunday meetings, Mueller also orchestrated the transfer of a large number of US Air Force officers under Maj Gen Sam Phillips, who brought to NASA the "systems engineering" approach which had previously been so successful in delivering ballistic missile programmes such as Thor, Atlas and Titan, fast and with great efficiency. Phillips was turned over to NASA on 30 December 1963.

In January 1964 Phillips brought with him 55 senior Air Force officers under Project APO, or "Project 55" as it was unofficially known. Never had there been such a high-power gathering of senior Air Force leaders seconded in "mufti" to get NASA restructured, reorganised and back on track: Brig Gen David Jones became deputy assistant to Phillips; Col C. Bolender became Apollo Mission Director; Col E. O'Connor became Director of MSFC Industrial Operations; and Col Sam Yarchin became Director of the Saturn V office. Added to which were new versions of PERT forcibly imposed by Phillips and a Configuration Control Board set up to standardise all the field centre working by formalising common projects.

Surprisingly, von Braun's team at Huntsville were still working management flow monitoring via wall-mounted waterfall charts but quickly came to see the advantages of the more sophisticated systems-management PERT routines. But it was a culture shock. Overcome too were the lethargic test programmes for qualifying new rockets. The first few flights of Saturn I had been conducted with just the first stage alone, engineers gathering as much data as they wanted before adding an upper stage and testing it with that.

Replacing this ponderous approach was Phillips' all-up systems testing philosophy whereby from the very first launch the entire rocket stack was assembled and fired sequentially. Phillips' team argued that only in this way could the fully interactive responses of a complete launch vehicle be fully understood, on the basis that a first stage alone will act structurally very differently to how it responds when upper stages are added. This all-up testing approach was a bold shift and von Braun argued long and hard against it but he

was overruled and quickly came to accept that as the best way forward.

It was this approach that would allow the very first flight of the mighty Saturn V rocket to fly on 4 November 1967 with all stages live and to send its unmanned Apollo spacecraft on a highly elliptical path out into space and back again, qualifying not only the rocket but the Apollo Command Module as well, returning at high speed to simulate a fast descent from the Moon.

As well as parallel systems development, where all elements of a rocket or spacecraft were developed and tested concurrently, Phillips' team developed a dramatic shift away from the conventional string-led fault analysis and failure mitigation sequence calculated in series, to a parallel system string analysis, an approach which did much to improve reliability and to perfect the translation of a systems engineering approach, elegant in theory, to one in which its practical application showed fewer failures and greatly improved reliability. It was this approach that carried NASA through some of its most traumatic episodes in manned space flight, including the management of the Apollo 13 crisis in April 1970.

It is difficult today to over-emphasise the value of these US Air Force officers, for they completely changed the NASA culture, giving it a military-style efficiency with tough approaches

BELOW In December 1968, on only the second manned Apollo flight, astronauts Borman, Lovell and Anders orbited the Moon, sending back the first views of the Earth taken by a hand-held camera around another body in space. *(NASA)*

OPPOSITE Neil Armstrong takes a photograph of Edwin Aldrin (who later changed his name to Buzz Aldrin) deploying a passive seismometer at the Apollo 11 landing site where the two astronauts had touched down on 20 July 1969, fulfilling Kennedy's instruction. (NASA)

to challenging management issues and complex technical solutions. NASA simply did not have the experience in giant programmes to carry it off and at peak the agency had more than 400 US Air Force officers on loan inserted into every facet of the Apollo programme. Jim Webb had been at the very core of this induction cycle and had personally received acquiescence from Robert McNamara, Secretary of Defense, to get these men released from high-profile career paths to work on a programme with which many of them had little affiliation. And, it has to be said, little inclination when promotional opportunities within their own service were frequently put aside to serve the nation in its drive to be first on the Moon.

The management changes and the revolutionary techniques and solutions provided by men like Mueller, Phillips and a host of other inductees created a different NASA, a fresh and revitalised agency which was still racing with uncertainty toward the Moon landings. While in 1963 President Kennedy had been talking up the possibility of either doing a deal with the Russians to "go to the Moon together" or even of seriously scaling back the goal, in 1964 there was a stoic determination to press on and fulfil the late President's challenge. As said, it was only after the assassination and realising the earnestness with which Apollo was picking up pace that the Russians decided to make it a race out of a solo run.

But at NASA, coming up to begin the Gemini flights, delayed beyond the projected schedule, there was a faltering uncertainty. Webb was desperately worried that Gemini flights would get in the way of Apollo, which he knew to be the late President's obsession and which he had personally signed up to, not so much out of agreement with the goal but as a member of that now rare breed who see civil service as just that – a service to the commander-in-chief. An edict from the highest office in the land had determined, on behalf of the American people, that this was the way forward and who was he to gainsay that! His species would become increasingly isolated and rendered almost extinct in the decades to come.

George Mueller saw Apollo more as a door to the human exploration of deep space and the broader environs of the solar system and he

received some encouragement in that direction from Lyndon B. Johnson, now President, who had masterminded the passage of the National Aeronautics and Space Act through Congress in 1958 and who, as Vice President, had steered Kennedy toward a Moon goal rather than some other objective. But concern at Gemini's flagging pace (more a result of the knock-on effect of the poorly constructed administrative control prior to 1963) prompted him to deliver a cut-off date of 1 January 1967 for the flight phase of Gemini. The last manned Gemini mission returned to Earth on 15 November 1966.

While the public were infatuated with the race to the Moon, NASA was achieving outstanding accomplishments with a successful fly-by of Mars in 1965, returning the first pictures of its barren surface, following a fly-by mission to Venus in 1962. It had also received funding to start development of two unmanned spacecraft on fly-by missions to Jupiter and Saturn. Meanwhile, Ranger impact probes had taken close-up photographs of the Moon before crashing and Surveyor spacecraft had landed on the surface while orbiters had begun mapping potential Apollo sites. But it was the human exploration of the Moon that gripped the imagination and excitement gathered pace with the ten successful Gemini missions.

The first Apollo flight was scheduled for 21 February 1967 but serious problems with both the spacecraft and the Saturn V rocket were being feverishly worked on by Mueller and Phillips, shielding Webb from poor performance by the spacecraft manufacturer North American Aviation who was similarly working the troublesome S-II second stage for the big Moon rocket. Moreover, a whistle-blower at NAA had warned of shoddy workmanship and poor quality control. Phillips had written to NAA warning of dire consequences and had been working to improve their performance.

On 27 January 1967 astronauts Grissom, White and Chaffee lost their lives in a fire that broke out in their Apollo 1 spacecraft during a simulated launch at Cape Canaveral. Although astronauts had lost their lives in car crashes and flying accidents, these were the first men to lose their lives in a spacecraft. Less than two months later Soviet cosmonaut Vladimir

Komarov lost his life returning from space in Soyuz 1, an accident caused by the haste to develop Russia's second-generation spacecraft. The race was taking its toll. But if these tragic disasters could perhaps have been avoided by less pace and a more measured effort, NASA's performance during the investigative process that followed would receive high praise and free it from a tighter scrutiny that would in reality only have delayed the programme.

From the moment the disaster struck, NASA plunged into an open-house approach, holding numerous press conferences, neither transferring nor apportioning blame and at Webb's deft handling the agency was allowed to conduct its own internal inquiry. A fast rising star of NASA diplomacy and trusted friend of many in both the executive and the legislature, for Congress and the White House, astronaut Frank Borman became the voice of NASA, with the added relevance of being an astronaut and from the fraternity that had lost three of its own. To the general public and the lay audience awaiting news, his calm observations were reassuring and confidence-building.

NASA came back from the Apollo fire stronger and more determined than ever to right the wrongs, pick up the pieces and get on with the job. Within 30 months of the fire two astronauts were walking around on the Moon and the space agency was at the pinnacle of its fame, a symbol of the "can-do" spirit and a shining example of a culture struggling to come to terms with war in South East Asia and burned-out cities hit by race riots across America. And it was largely these factors that doomed the space programme to a period of retraction and self-diagnosis as to how to best utilise the vast storehouse of technical innovation, engineering, scientific exploration and human endeavour that had characterised the first 12 years of NASA.

NASA had intended using Apollo as a springboard to further exploration, beyond the Moon and on to Mars. Since 1964 it had been conducting low-level studies on stretching the remaining hardware procured for the Apollo programme and using spacecraft and rockets surplus to the goal. Originally, NASA had ordered 20 combinations of Saturn V and Apollo spacecraft as well as other spacecraft for launch on the smaller Saturn IB. Initially, NASA used these combinations to plan a further nine flights following the successful mission of Apollo 11 in July 1969 put Armstrong and Aldrin on the lunar surface. The third attempt at a landing failed when Apollo 13 suffered a near-disaster and NASA only just managed to get the crew back alive.

With a significant decline in its budget and with little interest among the public in continuing with lunar landings, three were cancelled, leaving only four to complete the Apollo Moon series. The last three were extended in duration, significant upgrades having been applied to the Lunar Module to support three full working days on the surface with a Lunar Roving Vehicle carried to each site, the last of which was reached by Apollo 17 in December 1972.

NASA was able to use redundant hardware to mount three visits to a makeshift space station called Skylab in 1973–74 and to conduct a joint flight with the Russians when the last Apollo docked to a special module attached to a Soyuz spacecraft in 1975. Yet even before the first manned landing on the Moon, NASA had been looking at a radical new way of supporting a human space flight programme, although President Nixon, in the White House since January 1969, had briefly looked at abolishing NASA altogether. When Vice President (to Eisenhower), Nixon had never supported a permanent space programme and carried that ethic into the top office eight years later.

With a slashed budget and diminishing support from the White House, NASA sought ways to cut costs and retain its manned programme, seeking in a reusable Shuttle a cheaper way of riding out the lean times until better days returned and it could once again command sufficient resources to build a space station and reopen the possibility of carrying Americans to Mars. NASA had made outstanding progress by visiting every major planet in the solar system but it was unable to progress with its original vision of a permanent presence on the Moon and expeditions to Mars. All that would have to wait.

Support for the Shuttle was granted by President Nixon on 5 January 1972 but the original plan to develop a post-Apollo programme based on a reusable Earth-to-orbit shuttlecraft, a space tug for accessing and repairing satellites in space, a permanent space station, a nuclear shuttle for moving rapidly between Earth and distant points in space and a commitment to a manned Mars landing was scrapped by Nixon, only the Shuttle remaining as a jobs programme for California – Nixon's home state. This all but decimated NASA and not before 12 April 1981 did the first Shuttle carry astronauts into orbit.

A New Dawn

The transformation of NASA and its original paradigm of space station/lunar landings/Mars missions in that sequence had been torn up in May 1961 when President Kennedy snatched the mid-term goal and redirected NASA to an immediate, all-out effort to put astronauts on the Moon. Seeking to return to its original long-range plan, NASA had no money for the parallel development of both Shuttle and space station but the first few flights of the Shuttle convinced the White House to make a bold announcement. On 25 January 1984 a new national leadership drive begun by President Ronald Reagan relit the vision and authorised the space agency to develop a permanently manned space station and "to do it within a decade".

A tour by NASA Administrator Jim Beggs recruited the Europeans, the Japanese and the Canadians to join together in a partnership which would collectively provide the modules and the physical structures necessary to build what was at first called Space Station Freedom.

The world was still in the deep chill of a Cold War and Reagan liked the notion of using orbital flight as the flagpole for an ideological message. Congress was not quite so convinced and failed to provide the funds necessary to have Freedom operational by 1994. In fact events transpired to change the very nature of the station.

Shortly after the decision to build Space Station Freedom NASA suffered its second catastrophe when Space Shuttle *Challenger* exploded shortly after lift-off on 28 January 1986 when extremely low temperatures caused seals between solid rocket booster segments to leak, delivering a crippling blow to the national space policy. NASA had wanted to

LEFT The surge in public awe and adulation at humanity's first landing on another celestial body was reflected throughout the commercial world and even extended to sound recordings of the Moon landing released by Decca. *(Decca Records)*

LEFT A flagship programme for NASA, the two Viking missions to Mars were bold, in that they pioneered techniques from which engineers are still learning today, and ambitious in the breadth of their scientific tasks. The Orbiters remained in orbit about the planet mapping surface details for final landing site selection and continued to send back images for many years after they first arrived in June and August 1976. *(NASA)*

ABOVE **The two landers successfully reached the surface of Mars, in July and September 1976, respectively, continuing to operate for 3.6 years and 6.25 years each. During this time they survived several seasons and sampled the surface physically and chemically in an unsuccessful search for life.** *(NASA)*

BELOW **Each Viking Lander was powered by a radioisotope thermoelectric generator using the heat from a plutonium source to produce electricity and carried a biological analysis package in addition to a surface sampler and scoop which transferred soil to various instruments on board.** *(NASA)*

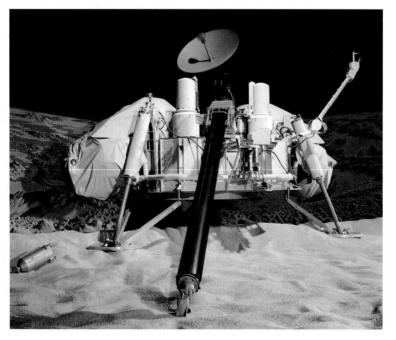

use the reusable Shuttle to replace all those old expendable launch vehicles, most converted from early ballistic missiles. It had run down production of the Delta class launcher until only two remained to be delivered as the Shuttle began to take over all the satellites previously launched by conventional rockets.

In the resulting inquiry it was decided that the Shuttle should not carry commercial satellites for fee-paying customers – a service that had pressured NASA to launch "on time, every time", according to NASA manned flight boss Gen James A. Abrahamson. This was a shift in national policy because the US launch industry was so entwined within the decisions made by the government about how it would use the Shuttle. But the inability to launch the Shuttle as frequently as had been imagined increased pressure not to succumb to the kind of temperature constraints which logically should have prevented the launch on the freezing January morning when *Challenger* was lost.

NASA went through another transformation as a result of *Challenger*, abandoning hope of using the Shuttle to replace expendable launch vehicles and restricted instead to carrying government payloads such as science satellites and some military payloads but primarily used for lifting into orbit all the various elements of the International Space Station, which at the time of the inquiry into the disaster was still some way off. The loss of *Challenger* held back many payloads and science satellites as well as observatories such as Hubble until the backlog could be accommodated and the expendable launcher industry remobilised. Paradoxically, the Air Force had never fully trusted the Shuttle as a truly reliable transportation system, despite having booked several satellites on it. But in the aftermath even the ISS was in peril.

When the dissolution of the Soviet Union began in the late 1980s and the communist system collapsed at the end of 1990 with the removal of Mikhail Gorbachev, it provided a new opportunity for cooperation with the new democratised Russian state which threw a lifeline to the station. President Bill Clinton offered the Russians a part of the space station programme, still floundering along with numerous design changes and cost-cutting plans to get it through Congress.

The Russians had been launching and operating space stations for 20 years, while America went to the Moon and re-engaged with the remit of the old von Braun paradigm. The Russians had a lot to offer and many at NASA remembered the cooperative joint flight of 1975 and recognised that much could be learned that would be of great advantage to the US-led station. The Russians were still operating the massive Mir space station complex and were reluctant to abandon it, bringing friction from NASA which wanted them to shut it down and focus exclusively on being a partner in the US-led project.

Moreover, with the White House manipulating the two players together to maximise the value to the US while offloading still further some of the cost of development and assembly, it made political, financial and technological sense to bring them in. Recognising that there was very little money for an indigenous human space flight programme, the Russians accepted and on 7 December 1993 – almost ten years after Reagan first ordered NASA to build a station – the roadmap was finally set to really make it happen. Now it was necessary to find another name and Freedom became the International Space Station (ISS), assembled with modules built in the USA, Europe, Russia and Japan with Canada providing the robotics – an apt choice since it had been responsible for the manipulator arm for the Shuttle.

This was the new age of international cooperation with a vengeance but it had been evolving for a long time. When NASA found it could not afford both a Shuttle and a space station, in the early 1970s it had wooed the Europeans into building Spacelab, a pressurised laboratory module to be carried inside the cargo bay of the Shuttle where astronauts could operate experiments. There was also a pallet assembly for carrying exposed experiments either exclusively or in addition to the habitable

SKYLAB

module. The agreement to build Spacelab was signed by the Europeans in August 1973 and the first pressurised module was carried by Shuttle *Columbia* launched on 28 November 1983 at the start of a long period of successful science missions providing NASA with the opportunity to use a mini-space station and European astronauts the chance to fly with NASA in the Shuttle.

The last Spacelab pressurised module was carried by *Columbia* on 17 April 1998 as

ABOVE Assembled from redundant Apollo hardware, Skylab was the world's first space station to receive multiple crew visits, three of which took place during Skylab operations in 1973 and 1974. *(NASA)*

SATURN WORKSHOP 1-G TRAINER

RIGHT Spacious and equipped with a wide range of instruments for studying the Earth, the Sun, astronomical objects and the effects of long-duration space flight on the human body, Skylab provided much needed experience in how to live and work in space for long periods. *(NASA)*

the ISS began to emerge but the station was proving a difficult project to manage, given the extensive international effort to bring it to reality. It took longer than expected to work out all the contractual arrangements and solve the technical difficulties that were essentially bringing two giant space programmes together. Where once they had been ideological adversaries, playing out roles defined for them by their respective political overlords and

leaders, scientists and engineers were now working out the difficulties together.

The first launch to build the station in space took off from Russia's Baikonur launch site, from where Sputnik 1 and later Yuri Gagarin had been launched, on 20 November 1998. It was followed by a steady succession of launches from Kourou in French Guiana where the European Space Agency (ESA) used its own Ariane V rocket to lift cargo modules, from Tanegashima where the Japanese did a similar job, and from Baikonur in Kazakhstan where Russia launched its own modules. From Cape Canaveral, NASA launched the pressurised modules for US components as well as those for ESA and the Japanese.

On through into the new century the giant assembly work progressed until on 1 February 2003 the Shuttle *Columbia*, returning from a science mission unrelated to the ISS, burned up on re-entry. Fragile thermal protection had been damaged during launch almost 16 days earlier when insulation from the external propellant tank broke free and struck the leading edge of the orbiter's left wing. Assembly of the ISS was held up during the stand-down until flights with the Shuttle resumed on 26 July 2005. The effect on America's space programme was greater than the loss of *Challenger* had been almost exactly 17 years earlier and would be lasting.

Since 2000, NASA had been looking at the Shuttle programme, evaluating whether to plan its retirement after the ISS was built or apply funds to a complete refurbishment of each orbiter, incorporating new and safer systems together with improved methods of escape, albeit limited as they were. The loss of *Challenger* had brought a new resolve to fund a replacement, the orbiter *Endeavour*, but also to preserve it for the less fast-paced routine of building the station and offload commercial satellite flights. But the loss of *Columbia* narrowed options and President George W. Bush decided to support retirement of the

Shuttle after the ISS was complete in 2004 and to divert NASA's attention to sending astronauts back to the Moon.

Since the early 1990s NASA's budget had been stable, showing very little sign of growing so every programme or long-range plan had to fit within a flat funding line. Gone were the days when NASA was working on four manned space vehicles simultaneously; now, it was not possible to start building a successor before the precursor had been retired. NASA organised the Constellation programme which required the construction of two new launch vehicles – Ares I and Ares V – and a manned spacecraft called Orion supporting the return of humans to the Moon by 2020 at the latest. It would be funded through savings from retirement of the Shuttle and from the US withdrawing from the International Space Station.

BELOW Space Shuttle *Atlantis* flies to a building site in space during June 2007 to deliver weighty truss assemblies and a set of solar arrays for the International Space Station. *(NASA)*

RIGHT Just some of the more than 30,000 thermal protection tiles which protected the Orbiter from the high temperatures of re-entry are visible across the forward and lower surfaces of *Atlantis*. *(NASA)*

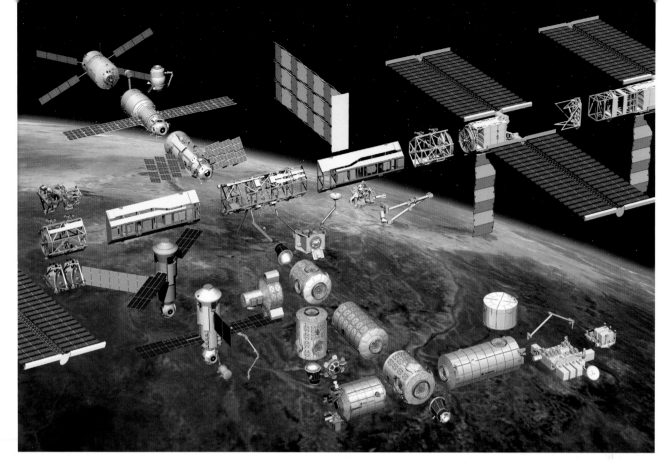

ABOVE The International Space Station has been one of the greatest successes of the modern era, a vast assembly of modules, trusses, solar arrays, thermal radiators, airlocks, viewing cupola supported by a continuing flow of commercial cargo carriers.

The designations for the two Ares rockets were notionally in recognition of the famous Saturn I and Saturn V rockets of the Apollo era but they signified more than that alone: Ares I would have a single solid rocket booster as the first stage with a liquid propellant upper stage; Ares V would have two Shuttle solid rocket boosters with five segments apiece flanking a first stage core being a lengthened variant of the Shuttle tank but with five RS-68 engines, a type developed for the Delta IV heavy lift launcher. Ares would have the capacity to place 414,000lb (188,000kg) in low Earth orbit or send 157,000lb (71,000kg) to the Moon compared with 99,000lb (45,000kg) for the Saturn V.

Concern over the actual cost of the Constellation programme, itself seen as a stepping stone to Mars, prompted President Obama to call for a commission to examine the funding profile, which judged that it could not be achieved within a fixed-level budget which the President was not persuaded to increase. On 1 February 2010 he announced that Constellation was cancelled but a rebellious Congress, outraged at the decision, voted for NASA to develop a new launch vehicle known as the Space Launch System (SLS) and to resume work on Orion. There was no defined

objective but the US would retain a heavy lift capacity and a deep-space exploration vehicle capable of carrying humans on a variety of ambitious missions.

Meanwhile, in an initiative begun under the Bush administration, NASA continued to pursue investment in commercial companies building systems capable of supplying the International Space Station with cargo and supplies after the Shuttle had been retired. Two primary contenders emerged: SpaceX and Orbital Sciences. NASA would use these contractors to provide a lifeline to the ISS, anticipating a commercial crew launch service later in the decade. But Congress feared duplication and proved reluctant to provide the necessary funds for rapid development of a commercial cargo capability, eventually voting for the necessary budget for that only when European and Japanese unmanned cargo supply ended.

The return of orbiter *Atlantis* in the pre-dawn hours of 21 July 2011 closed the operational phase of the Shuttle programme. Over a period of more than 30 years, 135 launches had taken place with a vehicle mistakenly assigned at its inception to lower mission costs and replace expendable rockets. From the first audacious flight on 12 April 1981, a mere 20 years after

Yuri Gagarin became the first human in space, the Shuttle proved challenging and more difficult to operate than anyone believed during its gestation.

The challenges faced by Shuttle engineers, technicians, managers and astronauts were unprecedented, with several technologies employed which had never been built when the programme started. Without precedent there was no historical base from which to learn lessons and without parallel there were no comparisons from which to make decisions about design and operability. It was truly unique in the full definition of that word and its type will probably never be seen again. Conceived in the height of the Cold War, approved even before the last Apollo missions had flown to the Moon, the Shuttle will always be a shining example of what is possible. But there had been a price.

The loss of two orbiters had taken 14 lives but the achievements of the programme were unique: launch and servicing of the Hubble Space Telescope; assembly of the International Space Station; base for numerous independent science missions with Europe's Spacelab which pioneered the way for cooperation on the ISS;

and not least the ability to bring together former adversaries in an enduring period of peaceful cooperation. This winged, reusable shuttlecraft, a concept dreamed of for decades, was finally realised through a national effort at least as challenging as the Apollo programme had been in its day.

NASA had matured through two sequential periods of enormous change and adaptation fuelled with triumphs and tribulation. When the Shuttle landed for the last time there was no clear objective for the future – only a very big rocket (the SLS) and an Apollo lookalike (Orion) which returned space exploration to the era of expendable rockets and ballistic capsules. Part of the problem was the dysfunctional relationship between President Obama and Congress, which chafed at pressure from the White House to cut NASA out of the human space flight business and rely on commercial carriers for sustained use of the ISS. Obama wanted to retain the ISS and use private contractors to keep it operating but was reluctant to give NASA a deep-space exploration objective which only big government programmes could accomplish.

BELOW The ISS complete and fully operational, supporting human occupation without pause to this day since 09.21hrs UTC on 2 November 2000.

Today and tomorrow

Since long before the formation of NASA, humans dreamed of travelling to Mars. That has become the coveted goal of space-faring nations around the world. Notwithstanding the dramatic change in understanding about Mars, now revealed as a seemingly barren world probably devoid of ever having supported life, it still beckons as a benchmark for technological achievement and satisfies the human urge to explore and seek the summit of the next hill. That desire has sustained Mars exploration using robotic spacecraft and has kept alive the hopes and dreams of advocates for a human presence on the Red Planet.

Until the inauguration of President Trump in January 2017 NASA was holding fast to a roadmap approved by the previous administration, which essentially had not decided upon a mission goal for the SLS/Orion programme instigated by Congress. Options included a rendezvous with an asteroid to retrieve samples for analysis back on Earth or a circumlunar mission to dock with an asteroid previously retrieved by an unmanned spacecraft and transferred to the vicinity of the Moon. To a degree these were seen as vanity missions accomplishing nothing other than brandishing the capabilities of the new heavy-lift/deep-space capability of the evolved hardware.

Early in 2018 President Trump approved a plan to extend the international partnership formed through the ISS programme and provide a meaningful goal for SLS/Orion by focusing on a set of sequential steps toward the long-term objective of getting astronauts to Mars. Agreement has been reached for Russia, ESA and Japan to join NASA in developing a mini-ISS in lunar orbit, assembly of which will begin in the mid-2020s. Orion will be the means of moving astronauts between Earth and this Lunar Orbital Platform-Gateway (LOP-G) which will incorporate elements provided by all the partners.

The plan is to use this platform to place astronauts down on the surface of the Moon and to also exploit its deep-space location to test and evaluate various systems essential to getting to Mars, not least an electric propulsion system capable of moving large payloads across the solar system. But the core value of the Gateway will be to have astronauts remain in lunar orbit for extended periods to study the physiological consequences on the human body of prolonged exposure to space and radiation outside the protective envelope of the Earth's magnetosphere.

Any trip to Mars will require humans to remain in this unprotected environment for up to two years and there is a complete lack of knowledge about how the body will respond to these conditions. The ISS is unlikely to remain active beyond the 2020s, although some commercial exploitation may be possible as private companies and corporations take over more of the fundamental support infrastructure for long-term maintenance of existing facilities. Such leasing arrangements began on Earth with NASA facilities when it first formed, extended into space through support for the ISS. NASA sees such a commercial support structure for maintaining the Gateway and eventually, several decades in the future, of maintaining a permanent presence on Mars.

But the accomplishments of the past 60 years and the achievements of the world's first civilian space agency have been outstanding as NASA extended its reach throughout the solar system. Ever since the first successful fly-by of Venus in 1962 and the first successful fly-by of Mars in 1965, NASA has led the way right out across the solar system, visiting Jupiter, Saturn, Uranus and Neptune and passing close by the dominant minor planet of the Kuiper Belt – Pluto – on 14 July 2015. NASA has successfully sent 17 spacecraft to Mars, of which three conducted fly-bys, seven went into orbit, three landed and four more wandered around on the surface.

Since October 2000, humans have had a sustained presence off Earth at the International Space Station; at no time since then has there not been at least two humans in orbit. Humans have begun to migrate from Earth to the near reaches of the solar system and eventually to a permanent presence on other worlds – first the Moon perhaps and then Mars. It is a capability which speaks to the very best in the human condition: groups working peacefully together, sharing a journey through time toward a distant goal of leaving this solar system behind and starting a migration to other planetary systems in the galaxy. And NASA will be there – all the way.

MEATBALL OR WORM?

Everyone knows the NASA badge, more properly known as the "seal", but who knows how it came to be that design and who remembers the catastrophic decision to change it, before reverting to its original form under a wave of protest and public pressure? The story and the explanation tracks shifting trends in design as well as marking cultural changes on society at large.

James Modarelli departed from his paid job at the Lewis Research Center Reports Division in 1959 to design what has become one of the most iconic emblems for any government department. Affectionately known as the "meatball", it combines visual metaphors where the sphere represents a planet and the star represents space. The red chevron represents flight and is shaped in the manner of a supersonic aerodynamic profile while the orbiting satellite represents an elliptical path around the heavens. It was conservative as befitted the time, NASA having metamorphosed from the NACA.

In 1974, responding to the Federal Graphics Improvement Program of the National Endowment for the Arts, NASA had the mistaken idea that the old seal was outdated. It was the end of an era, thought some. The last Apollo spacecraft was about to fly to link up with a Russian Soyuz spacecraft and for two years industry had been working to build the world's first reusable transportation system, the Space Shuttle. Surely this was the time to upload NASA to a new age of space.

The agency paid Richard Danne and Bruce Blackburn to come up with a new logo, one fit for the new era. The "worm" was the result, the letters in red and the "A" for aeronautics reduced to a pseudo-wigwam shape. There was consternation at headquarters and senior managers took opposing views – some liking it, others hating it. However, the custodians of art and culture (presumably) liked it and the design received the highly coveted Award of Design Excellence from The Presidential Design Awards. The general public were divided too, but unequally, most opting to support the "hate it" lobby. And so began a campaign to get back the meatball.

It might have suited the elevated halls

of good taste and fashion, emblematic of the 1970s, but most said that this was the defining reason for rejecting it. But having paid taxpayers' money to change all the stationery and decals applied to structures, buildings, rockets and spacecraft, it held fast. Even through the 200th anniversary celebrations in 1976 commemorating the founding of the United States. Eventually, the naysayers had their way and in 1992 NASA reverted to the meatball. Like it or not, this timeless rendition of NASA's goals embracing both aviation and space says more than any worm ever could!

LEFT The original NASA logo representing the two themes of aeronautics research and space exploration as enshrined in the National Aeronautics and Space Act of 1958. *(NASA)*

LEFT For a brief period during the 1970s, the "worm" took over but widespread indignation soon saw its replacement with a more traditional version of the classic logo. *(NASA)*

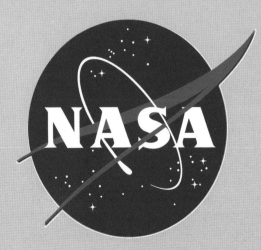

LEFT The current logo, brought up to date but bearing the same iconic characteristics. *(NASA)*

HUMAN SPACEFLIGHT
A NEW FRONTIER

NASA

...ruce McCandless, a few feet
...m the space shuttle Challenger,
...er untethered spacewalk,

Please do not touch
the exhibit.

Thank you!

...not touch
...e exhibit.

...Thank you!

HUMAN SPACEFLIGHT
A NEW FRONTIER

The Cold War between the United States and former Soviet Union gave birth to the Moon race and an unprecedented program of scientific exploration. The Soviets sent Sputnik, the first artificial satellite, into orbit on Oct. 4, 1957, and the first person into space on April 12, 1961. In response, President John F. Kennedy challenged our nation "to achieving the goal, before this decade is out, of landing a man on the moon and returning him safely to earth." On July 20, 1969, people around the world witnessed and celebrated Apollo 11 astronaut Neil Armstrong planting the first footstep on the lunar surface.

In the 1970s, U.S.-Soviet political tensions that had accelerated the space race began to thaw. Competition gave way to cooperation between the two nations with the Apollo-Soyuz Test Project. International collaboration among many nations would become the norm during the space shuttle era and the ongoing cooperation with the International Space Station. These partnerships have taught us more about the universe, improved our lives at home, and expanded the possibilities for future exploration into deep space. The race to the Moon happened in three phases over a 17-year period. They were Mercury, Gemini and Apollo. Each program's set of goals led up to lunar exploration.

The Mercury Program

Project Mercury, the United States' first human-in-space program, made 25 flights, six of which carried astronauts between 1961 and 1963. The objectives of the program were to orbit a human space craft around Earth, to investigate a person's ability to function in space, and to recover both the astronaut and spacecraft safely. More than 2 million people from government agencies and the aerospace industry combined their skill, initiative, and experience to make the project possible. Mercury allowed that humans could function for periods up to 34 hours of weightless flight.

Mercury Astronauts

Mercury astronauts, the "Original Seven," Jan. 1959. NASA introduced its first astronaut class. Front row, left to right: Wally Schirra, Deke Slayton, John Glenn, Scott Carpenter; back row: Alan Shepard, Gus Grissom, and Gordon Cooper.

Freedom 7

Liftoff of astronaut Alan Shepard's Freedom 7 mission, powered by a Redstone rocket, May 5, 1961. Shepard became the first American in space, a flight that lasted 15 minutes, 28 seconds. He later made 6

Katherine Johnson

NASA research mathematician Katherine Johnson did the trajectory analysis for Alan Shepard's historic mission. Johnson worked at NASA's Langley Research Center from 1953 to 1986. She and many other women made critical technical

Mission Control

Mercury Mission Control, Flight Control Area. During Project Mercury, the front wall of the Flight Control Area featured a large world map display with the path to be followed by the capsule. A circle marked each station in the

Astronaut John Glenn

John Glenn orbited the "Friendship 7" Mercury spacecraft, Feb. 20, 1962. Glenn made history by becoming the first U.S. astronaut to orbit Earth.

2 NASA Field Centres

This section lists the various field centres, some of which have been completely transformed by changes in assignments, in the programmes they manage and in the roles and responsibilities they have been charged to carry out. Many are famous names, others less well known and some probably unfamiliar to all but the most ardent space enthusiasts. They are all here and are identified by the field name they have at the time of writing.

OPPOSITE The public have judged NASA as one of the most important government agencies for maintaining America's technological leadership in the world and the government itself has measured NASA as the most popular government body to work for. *(NASA)*

environment. Formed to mobilise the United States on new paths beyond Earth, it has grown and evolved over time as events have dictated the pace and the direction it has taken. When NASA began operations on 1 October 1958 it inherited a few existing facilities and a relatively small budget. It would grow far beyond the expectations of its founding leaders and would become one of the most famous government organisations from any country anywhere in the world.

The expansion of NASA in the first decade of its formation is astounding, increasing the acreage of all its facilities from 5,179 (2,095ha) to 142,000 (57,465ha) by June 1968 and expanding the total work area from 5,071ft^2 (471.1m^2) to 31,040ft^2 (2,883.7m^2). The total invested property value increased from $268 million to $4.4 billion in then-year dollars. In that first decade, four new field installations were built from the ground up (Goddard Space Flight Center, the Manned Spacecraft Center, the Kennedy Space Center and the Electronics Research Center). Because so much of

ABOVE 2018 marks the 60th anniversary of NASA and celebrations were held at all the field centres across the country. *(NASA)*

From its formation during 1958 the US national space agency has maintained a range of facilities dedicated to the exploration and exploitation of space and the deep-space

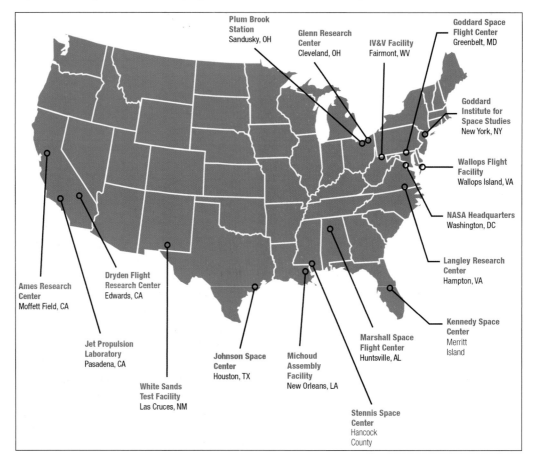

RIGHT Building on the original NACA formed in 1915, NASA has grown its capabilities through a wide network of field centres and facilities across the United States. *(NASA)*

Plum Brook Station
Sandusky, OH

Glenn Research Center
Cleveland, OH

IV&V Facility
Fairmont, WV

Goddard Space Flight Center
Greenbelt, MD

Goddard Institute for Space Studies
New York, NY

Wallops Flight Facility
Wallops Island, VA

NASA Headquarters
Washington, DC

Langley Research Center
Hampton, VA

Kennedy Space Center
Merritt Island

Marshall Space Flight Center
Huntsville, AL

Michoud Assembly Facility
New Orleans, LA

Johnson Space Center
Houston, TX

White Sands Test Facility
Las Cruces, NM

Jet Propulsion Laboratory
Pasadena, CA

Dryden Flight Research Center
Edwards, CA

Ames Research Center
Moffett Field, CA

Stennis Space Center
Hancock County

NASA's work was contracted out, the value to the industrial base of the United States was phenomenal. During the first ten years, it installed capital equipment then worth $400million in 20,000 prime and subcontractor facilities.

Over time NASA has acquired, or temporarily operated, several unique offices or small facilities which have long since been passed on to other users. One such opened in July 1962 as the Eastern Operations Office and which was to be the site of the Electronics Research Center (ERC), formally in existence as such from 1 September 1964. It was set up as a graduate training facility but its primary function was to develop an expertise in electronics during the rapidly expanding challenges of the Apollo programme. Situated in Cambridge, Massachusetts, opposite the main MIT building, the ERC was developed to a level where NASA expected it to have a staff of 2,100, of which 1,600 would be professional workers. Its importance was equal to that of the Langley and Marshall facilities.

The story of the ERC is too convoluted and structured by high politics and machinations to explore here and its history and eventual demise must be sought elsewhere but it is historically interesting in that it is the only NASA facility to have been closed. This despite its survival through one of the most troubled periods at NASA, when the collapse of NASA's budget saw layoffs and cutbacks at every other centre. The ERC was arguably one of the most important facilities, strongly supported by Administrator Webb but brought down by political infighting and partisan issues. Sadly, it was closed in June 1970 and the building is now the John A. Volpe National Transportation Systems Center.

But they are all here and provide a broad history of NASA and a record of its many programmes – some very well known and others hardly known at all. Each slots in to a specific reason for its existence while some have shifted roles and responsibilities over time. All of them have a proud history and an important story to tell. Due to the limitation of pages we can only do them scant justice, for to tell the full story of each would require many volumes.

Langley Research Center

When NASA began this facility was known as the Langley Aeronautical Laboratory, the oldest, and then the largest, of the NACA's facilities. Located about 150 miles (241km) south of Washington, DC, it employed 3,200 people and cost $126million a year to operate. Traditionally, its function was to conduct research into aerodynamics, aircraft structures and the operating problems of aircraft and objects travelling through the upper atmosphere and back to Earth, with about 40% of its work assigned to space research of various kinds, although in 1958 that term was rather loosely applied, much of that work being on understanding the upper atmosphere insofar as it related to flight.

In 1958 the LRC set up the Space Task Group (STG) under Robert Gilruth specifically to manage the MISS programme taken over from the Air Force and to coordinate facility activity on NASA's newly inherited manned space flight programme. This grew out of the Pilotless Aircraft Research Division (PARD) which had been set up just after the war to look into rocketry and its potential for human flight. But PARD had already got its feet wet in rocketry with a decision in 1956 to develop a multi-stage solid propellant rocket which would mature into the Scout launcher.

Several rebellious figures at Langley, a small, tight-lipped group mobilised by Max Faget, Joseph G. Thibodaux Jr, Robert O. Piland and William E. Stoney Jr, had thoughts of upstaging the Vanguard team for being the first to put a satellite in orbit to support the International Geophysical Year. After Sputnik 1 and the success of von Braun's Juno launch with Explorer 1 and Vanguard coming good, authorisation was received for development of the Scout launcher. The first launch occurred on 1 July 1960 and although that was a failure, the second attempt on 4 October 1960 was a success. Scout was transferred to Goddard Space Flight Center on 1 January 1991. It would achieve outstanding results before it was finally retired in May 1994 after 94 orbital launches.

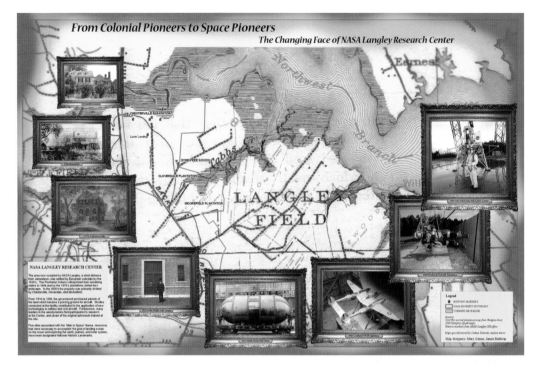

From Colonial Pioneers to Space Pioneers

The Changing Face of NASA Langley Research Center

Engineers at Langley were responsible for several vital steps in human space flight, the technical development of spacecraft capable of carrying astronauts into space, in pioneering the options NASA had for landing men on the Moon and the methods it would use underpinning lunar landing flights in the Apollo programme. Switching rapidly from working on development of the Mercury programme, Langley engineers decided to go all out for Moon landing studies and taught themselves the mathematics of orbital mechanics using a book written by British author Forrest R. Moulton.

Especially enthusiastic, engineer Clinton E. Brown capitalised on a fascination with Moon trajectories which had begun before the Apollo challenge was laid down. While received wisdom taught that planetary flights would begin first with orbital missions, then manned space stations followed by Moon flights, Brown wanted to go straight from Earth orbital tests to the lunar surface and set about planning how best to achieve that. As early as February 1959, Brown and his team set up a working group to lobby for this goal more than two years before Kennedy would select it. With Robert "Bob" Jastrow as its chair, the group spun off a Lunar Exploration Working Group under Brown and the lobbying team was all set to promote this approach.

A lot of things would have to happen before it could become a formal goal, not least a change of President and the shock news of Yuri Gagarin but Brown's studies were the very core of recommendations which would prove pivotal in giving NASA leadership the confidence to recommend a lunar landing goal when Vice President Lyndon Johnson went shopping for ideas on how to beat the Russians.

Langley was pivotal in getting men to the Moon in so many influential ways. In addition to the drive of Max Faget's people, and the

RIGHT The lunar orbit and landing approach simulator at Langley assisted Apollo astronauts with developing control systems for manoeuvring spacecraft close to the Moon. *(NASA)*

work of Clinton Brown, John C. Houbolt would pioneer the Lunar Orbit Rendezvous mode which was finally adopted in late 1962 and which would be the way Apollo evolved. After the formal announcement of the Moon landing goal, Langley bequeathed much of its relevant personnel to the new Manned Spacecraft Center but it focused too on how to control a manned vehicle in the final 150ft (45.7m) down to the lunar surface.

This resulted in construction of the Lunar Landing Research Facility consisting of a 400ft (120m) long A-frame gantry 250ft (76.2m) tall from which an overhead bridge crane would support a Lunar Excursion Module Simulator (LEMS) to give astronauts experience with "flying" a Lunar Module down to the Moon. When the Moon goal had been achieved, the facility was renamed the Impact Dynamics Research Facility and used to carry out crash tests on aircraft with a view to improving survivability. It is now known as the Landing and Impact Research Facility and is being used to conduct water splashdown tests with the Orion spacecraft.

Langley made additional contributions to the human space flight programme, including the six-degree-of-freedom Rendezvous Docking Simulator, completed in 1963 and used to give astronauts experience with manually docking a Gemini spacecraft to an Agena stage and later docking an Apollo spacecraft with a Lunar Module. Located in Building 1244, it provided an enormous volume of 210ft (64m) horizontally, by 16ft (4.8m) laterally by 45ft (13.7m) vertically for a three-dimensional simulation using full-size mock-ups of the forward section of respective spacecraft. The support structure was a three-axis gimbal frame suspended by eight cables from an overhead carriage/dolly system linked to an analogue computer.

After the decision to go to the Moon by Lunar Orbit Rendezvous had been made, the development activity for the spacecraft shifted to the Manned Spacecraft Center (the current

Johnson Space Center) with the production of the Saturn V undertaken by the Marshall Space Flight Center and flight operations at the Kennedy Space Center. But Langley pitched in to concerted efforts at developing space station concepts and led the field, in 1963 focusing on its Manned Orbital Research Laboratory (MORL) concept for which Boeing and Douglas had study contracts. But there was no money in the NASA budget to move productively toward a space station and that would have to wait until the Moon programme was over.

BELOW The drop-test rig at Langley's Lunar Landing Research Facility was employed in a variety of configurations to simulate the final approach phase of a descent to the surface of the Moon. *(NASA)*

LEFT Alan Shepard, America's first astronaut and Commander of Apollo 14, the fifth man to walk on the Moon, poses with the Lunar Landing Research Facility. *(NASA)*

With more immediate requirements at hand, however, Langley took an early lead in developing a spacecraft for carrying out photographic reconnaissance of the lunar surface to inform Apollo site selection teams about surface conditions down to a fine scale resolution. Under the Lunar Orbiter programme, Langley developed the first lunar research programme which demonstrated 100% success. Following on from the Ranger series of impacting photographic probes, in which only three of the nine launched were a success, and

BELOW A panoramic view of Langley's subsonic 14 x 22ft (4.26 x 6.7m) wind tunnel. *(NASA)*

BOTTOM An expansive view of the Langley materials research laboratory which is used for a wide range of tests and development programmes. *(NASA)*

the seven soft-landing Surveyor spacecraft, of which five were successful, five Lunar Orbiter missions were flown in 1966 and 1967 without a single failure. The images they returned constitute some of the best available image libraries available today and remain a strong source of reference.

Beyond human space flight and the exploration of the Moon with unmanned spacecraft, Langley has an extraordinarily proud record in the unmanned exploration of Mars, most notable for its work project managing the Viking missions to Mars launched in 1975 and which successfully put two spacecraft in orbit and two landers on the surface. It began when Langley introduced the idea of a very large spacecraft called Voyager (not to be confused with the outer-planet missions) in 1960 launched on a giant Saturn V.

After Voyager was cancelled, Langley got the project management lead in 1968 for a slimmed-down version, not least because of its success in managing the Lunar Orbiter missions, developing Viking with their outstanding strengths in technology development and the work on re-entry dynamics which would be valuable in understanding the problems effecting a controlled landing on Mars.

By the early 1980s Langley had largely returned to its origins – conducting vital and original research work on aerodynamics, on flight and the challenges of flying new and innovative aircraft. Manned flight was very firmly in the hands of the Manned Spacecraft Center and planetary missions were overwhelmingly in the hands of JPL.

Today, this facility is not known for that pioneering space work but without it none of the outstanding accomplishments achieved by NASA in human space flight could have been implemented. Moreover, it is supporting the next generation of launch vehicles with wind tunnel tests on the Space Launch System and a wide variety of scientific and engineering support work to assist NASA-wide programmes as well as contracted tests and evaluations on space science and engineering projects from industry.

Langley has about 1,800 employees and it is in the critical areas of aeronautical research that it is now dedicated, as well as advanced materials research, electron beam freeform fabrication and the application of plastics to reformation processes.

LEFT Langley operates a number of NASA aircraft on flight test, research and development programmes in close liaison with the Armstrong Flight Research Center. (NASA)

One of the original NACA facilities, in 1958 when NASA formed it was known as the Ames Aeronautical Laboratory. It employed 1,450 people and had an annual budget of around $87million. It supported a wide range of multifunctional facilities with a greater emphasis on high-speed aerodynamics, with a self-assessed 29% proportion of its work going on space activities.

Ames had been authorised by Congress on 9 August 1939 as the NACA's second laboratory, its location at Moffett Field, Mountain View, Santa Clara County, California selected from 54 candidate sites examined by a committee headed by Col Charles A. Lindbergh – the first man to fly solo across the Atlantic Ocean. The facility originated as an airship base for the US Navy prior to its transfer to the Army in October 1935 for use as a training base. It was reacquired by the Navy in April 1942 to become the Moffett Naval Air Station, with ground broken for the first NACA building on site on 20 December 1939.

The original name of the facility, Ames Aeronautical Laboratory, had been proposed by Dr Edward P. Warner to honour Dr Joseph M. Ames who had been chairman of the NACA from 1927 to 1939. The first 7ft x 10ft (2.13m

BELOW Ames Research Center, second of the NACA field sites, has majored on aeronautical research but after it was absorbed by NASA it established an outstanding reputation for flagship missions. *(NASA)*

x 3.05m) wind tunnel started operation on 13 March 1941 followed by the 16ft (4.8m) tunnel, the latter converted to a 14ft (4.26m) tunnel in 1955. Formal dedication took place on 8 June 1944 when the 40ft x 80ft (12.19m x 24.38m) tunnel was completed. Its present name was given when the NACA became NASA on 1 October 1958.

The mix of wind tunnels continued to grow with the Unitary Plan Wind Tunnel completed in 1956, which became the most commonly sought facility in the nation. It was built because no single tunnel could satisfy demands from a wide range of sources and the Ames tunnel complex had three separate test areas which were powered from a single centralised source. The transonic section measured 11ft (3.3m) by 11ft, and two supersonic sections were each 9ft (2.7m) by 7ft (2.1m) and 8ft (2.4m) by 7ft (2.1m). These facilities were quickly utilised in the development of airliners such as the DC-8, -9 and -10 aircraft, with military aircraft such as the F-111 fighter-bomber, the B-1 Lancer strategic and tactical bomber and the C-5A Galaxy heavy-lift transport aircraft.

One of the most ambitious, and useful, of all the tunnels at Ames is the 80ft (24.38m) by 120ft (36.57m) tunnel, the largest in the world. Entire aircraft, up to the size of a Boeing 737, are tested there along with a wide range of space vehicles and parachute devices for planetary spacecraft and although decommissioned by NASA in 2003 it is now under the management of the Air Force Arnold Engineering Development Center.

Support for impact analyses relevant to scientific analysis of objects raining down on the Moon and planetary bodies employed the Ames Vertical Gun Range, which provides support for geophysical programmes inside and outside NASA, the US Geological Survey being a sometime customer. The chamber is 8.2ft (2.5m) in diameter and can contain a wide variety of shapes and structures held at near vacuum states to simulate space environments. The chamber is also useful for carrying out research on penetrations of space structures, for spacecraft or space station modules.

NASA-Ames began to orientate itself toward life sciences in February 1961 and its first space project was the Pioneer programme from 9 November 1962, followed by the Biosatellite project on 13 February 1963. While these programmes would achieve public acclaim and recognition of the work at Ames by the media, the real substance of its work was elsewhere, sustaining an active programme of research into all forms of powered and unpowered flight. It had been the Ames aerodynamicist H. Julian Allen who, in 1952, had developed the blunt-body concept solving severe aerodynamic challenges about how to get missile warheads back from space.

This research was fundamental and crucial to the viability of the American nuclear deterrent, allowing re-entry bodies to make it safely back through the atmosphere at velocities which would cause the kinetic energy to produce temperatures beyond the ability of any known material to survive. Allen defined the solution by orientating the re-entry body so that it presented a flat (blunt) face to the direction of travel, creating a shock wave ahead of the object and standing free, ahead of the descending vehicle, rather than allowing it to make contact with the surface and conduct metal-melting temperatures to the structure. It solved the re-entry problem and has been adopted ever since for anything returning from orbit or beyond.

This work led directly to the study of conical "lifting bodies", blunt-shaped structures with flat or highly convex undersurfaces equipped with or without small lifting surfaces swept upward at acute angles from either side of the conical body for aerodynamic control. It was this work which led most industry proposals for the Apollo spacecraft to show such a configuration because there had been so many papers published about these shapes, with varying degrees of lift-over-drag (L/D). However, Apollo took the shape of a pure cone but lifting bodies were built and tested during the 1960s and it was this work that contributed so much to the design of reusable spaceplanes and shuttle

RIGHT Originally built as an airship shed, hangar No 1 is a visual icon of the compact facility in a beautiful and aesthetic setting. *(NASA)*

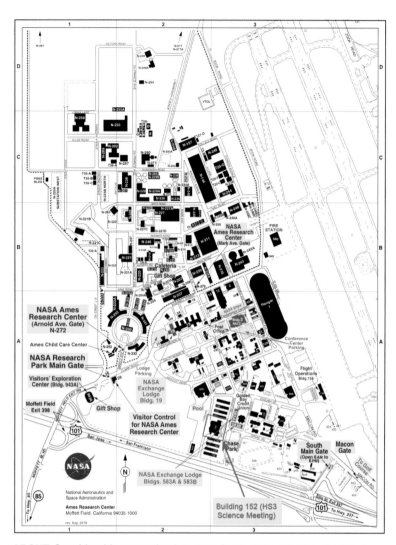

ABOVE Considerable expansion has seen Ames grow over the years but the availability of unique wind tunnel facilities has engaged in several cooperative projects with the Langley and Armstrong Flight Research Centers. *(NASA)*

LEFT The 80 x 120ft (24.38 x 36.57m) Ames tunnel is the largest in the world and is frequently used to gather flow and separation data on full-size aircraft up to small airliners. *(NASA)*

CENTRE The largest of the Ames tunnels seems even more spacious, a barn-like size which impresses the visitor on more than the first encounter! *(NASA)*

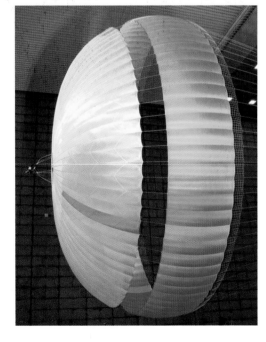

RIGHT Here employed for parachute retardation and analysis, the big wind tunnel is given scale by the 51ft (15.5m) diameter parachute employed by Mars Science Laboratory and the Curiosity lander and also for the Mars 2020 lander. *(NASA)*

vehicles that would materialise as the NASA Space Shuttle in the 1970s.

Ames used models fired from light-gas guns into high-speed jets flowing in the opposite direction to simulate high-speed atmospheric entry or used stationary models in gases heated by pebble beds, by rapid compression, or by electric arcs. Ames made a lasting contribution to aerothermodynamic research in arc-jet development initiated in 1956 and which took advantage of the additional funding and resources in 1961–62 to continue with this research and provide a wide range of data directly supporting the entry of probes fired into planetary atmospheres at great speed, surviving to transmit data.

Investment in Ames had been sustained from the time it was inaugurated, with additional wind tunnels expanding the capabilities of the facility. A 12ft (3.65m) wind tunnel was added in 1946 to investigate the aerodynamics of subsonic air vehicles in air streams with exceptionally low turbulence, together with a 1ft x 3ft (0.3m x 0.9m) supersonic wind tunnel in the same year. The need to study vehicle aerodynamics at hypersonic and supersonic speeds introduced a 6ft x 6ft (1.83m x 1.83m) tunnel in 1948 followed by a shock tunnel for studying space vehicle aerodynamics and heat transfer on re-entry vehicles.

Fundamentally, Ames became the go-to facility for the physical and life sciences and its scientists would play a highly significant role in both fields, including the development of biological instruments and experiments aligned with NASA's Viking Mars programme of the mid-1970s, which successfully put two landers on the surface equipped with instruments for detecting life in samples collected from the surface by a lazy-tongs device. Prior to this, it had conducted

eight successful Pioneer missions between 1965 and 1978, culminating in the Pioneer 10 and 11 missions to Jupiter and Saturn.

Exploiting the experience of these previous missions, Ames got the Lunar Prospector programme and on 6 January 1998 it was launched to the Moon, from where it carried out scientific research from polar orbit for a year.

But pioneering work is the order of the day at Ames, which since May 2010 has been flying a converted Boeing 747SP under a collaborative research programme with the German Aerospace Centre (DLR) observing magnetic fields in the galaxy, studying star formation and observing the centre of the Milky Way through an 8.2ft (2.5m) telescope with an oversize mirror of 8.9ft (2.7m) diameter for infrared observations. Known as the Stratospheric Observatory for Infrared Astronomy, SOFIA sustains a long tradition of using aircraft to reach high altitude where telescopic observations can be made in a less dense region of the atmosphere.

NASA-Ames is also engaged in cutting-edge development research of advanced supercomputing and the development of artificial intelligence (AI). These projects feed into the International Space Station, the various NASA space science missions, development of the Orion spacecraft and of the NASA human factors programme which underpins physiological research into all aspects of human space flight. It operates Pleiades, one of the world's fastest super-computers which achieved 10 petaflops of processing power in 2012.

The NASA Ames Exploration Center is located at the entrance gate to the facility and contains several iconic displays including a Moon rocket brought back to Earth by the crew of Apollo 15, the Mercury 1A spacecraft from 1960, and an immersive theatre with a 14ft x 36ft (4.26 x 10.97m) screen which has a rolling series of movies from major NASA missions.

ABOVE A specially converted Boeing 747SP provides Ames with a unique airborne research platform studying astrophysical phenomena. *(NASA)*

BELOW The Ames Visitor Center contains several artefacts and exhibits as well as a full-scale mock-up of an International Space Station module. *(NASA)*

The third of the NACA's original laboratories, this facility has got through more name changes than most. Congress had authorised a new facility for flight research on 26 June 1940 and 72 locations at 62 cities were examined as options. On 23 January 1941 ground was broken on a 199.7 acre (80.8ha) site adjacent to Cleveland-Hopkins Municipal Airport at Cleveland, Ohio. At first named the Aircraft Engine Research Laboratory, on 28 September 1948 it was renamed the Lewis Flight Propulsion Laboratory to honour Dr George W. Lewis (1882–1948) who had been the NACA's Director of Aeronautical Research from 1919 to 1947.

Activity had increased measurably during the Second World War with improved engines, lubricants, fuels and superchargers plus all manner of engine components. Quickly, after the war ended, work focused on jet engines, afterburners and combustion efficiency for the new age of jet aircraft. By the early 1950s rocket propulsion was added to the portfolio of research carried out at this facility and this resulted in some pioneering work on fluorine and hydrogen instead of using liquid oxygen as an oxidiser. Several studies examined the possibility of using a fluorine/hydrogen combination but these were never realised

BELOW Opened by the NACA as its third facility when it was known as the Aircraft Engine Research Laboratory, then the Lewis Flight Propulsion Laboratory, the Glenn Research Center is the home for aircraft engine research and rocket motor development. (NASA-GRC)

due to the difficulties with handling this highly reactive and toxic propellant.

With the formation of NASA in 1958 came another name change to the Lewis Research Center, operated with 2,700 employees and an annual budget of $120million a year. While it majored on propulsion and various powerplants for aircraft, 36% of work was attributed to space research. Focus switched to new ways to propel vehicles in space and to reduce journey times to distant places. This led to original work on ion propulsion and this was supported by the development of new facilities. By 1961, Lewis had the world's first laboratory-model mercury bombardment ion engine.

With increasing influence in the world of aviation and space propulsion, on 30 September 1962 Lewis got development responsibility of the cryogenic Centaur stage from the Marshall Space Flight Center followed by management of the Agena upper stage programme from 12 December 1962. Both Centaur and Agena would play seminal and enabling roles in launch vehicle technology and significantly expand the capabilities of each. Centaur would be the means by which Surveyor unmanned Moon probes were sent to the lunar surface and Agena, already an

LEFT The Mechanical Vibration Facility (MVF) test bed in its holding fixture. *(NASA-GRC)*

established upper stage, would secure many more development paths.

With Agena came responsibility for developing further derivatives of the Thor and Atlas first-stage elements and because it had demonstrated exceptional competence there, management of the M-1 rocket motor was shifted from Marshall to Lewis in October 1962. On 10 September 1964 project management of the 260in (660cm) solid propellant motor went to Lewis as well. Finally, on 1 April 1966, Lewis got control of the RL-10 engine which became an installed favourite for several launchers, not least the current Space Launch System.

Possession of the M-1 project was in several respects the peak of Lewis' engine development portfolio, a cryogenic LOX/LH2 motor with tremendous growth potential, designed initially for a thrust of 1.2million lb (5,337.6kN) but upgraded to 1.5million (6,672kN), then to 1.8million lb (8,006kN) and finally to a growth with reduced specific impulse of 2million lb (8,896kN). M-1 was considered as a second stage for an upgraded Saturn V, replacing the five J-2 second-stage engines which had a stage thrust of 1million lb (4,448kN) but its real potential would have been for a post-Saturn launcher. When that possibility was lost in the general mid-1960s run-down, the M-1 was cancelled on 24 August 1965.

By this date Lewis had begun to consider alternatives and NASA had already contracted several manufacturers, including Boeing, for post-Saturn rockets known as Saturn II. This work prompted a look at alternatives to the M-1 and the HG-3 motor was designed but never built. But the HG-3 was based around a high-pressure cryogenic engine which would emerge later as the Space Shuttle Main Engine built by Rocketdyne and improved by Aerojet General (which had initially lost the development contract).

Considerable development work was carried out on nuclear propulsion, with potential applications through the Saturn V launcher which could have considerably increased payload capability to deep-space destinations. NASA's Marshall Space Flight Center was keen to advance the development of the nuclear rocket programme, seeing in this technology an outstanding performance improvement for existing heavy lift vehicles. Lewis got management of the Nuclear Engine for Rocket Vehicle Applications (NERVA) programme which

BELOW The MVF with the table installed upon which instrument modules, satellites and spacecraft can be attached. *(NASA-GRC)*

would base its tests at Jackass Flats, Nevada.

Nuclear propulsion can come in many forms but in its simplest application it dispenses with the need for an oxidiser, since the energy to provide a strong reactive force comes from the thermal energy of a nuclear reactor to raise the temperature of the fuel to that usually achieved by chemical combustion. In NERVA, liquid hydrogen is the only propellant, delivered to a reactor and thence to the equivalent of a combustion chamber for exhaustion from a conventional convergent/divergent expansion chamber, or nozzle.

Managed by the Space Nuclear Propulsion Office at NASA Headquarters, NERVA was a cooperative venture with the Atomic Energy Commission (AEC) and was a product of work that had been underway since before the dawn of the Space Age. Early theoreticians in nuclear propulsion – as well as nuclear power for electrical systems on space stations – include Dr Leslie R. Shepherd of the British Interplanetary Society who worked extensively in the UK's nuclear power industry and from 1949 saw the advantages for space projects when he co-authored a paper on the subject with Val Cleaver, a rocket engineer and another Fellow of the BIS.

The NERVA programme fell victim to a retraction in NASA funding, the evaporation of plans for post-Apollo deep-space human space flight, where the nuclear rocket motor would excel, and it was eventually cancelled altogether in 1972. But Lewis had been engaged in power production for space vehicles over many years and responsibility for the power production system for the International Space Station fell to this facility when it was approved for development in the early 1980s. It was an apt choice, since the Center had been responsible for extensive photovoltaic research and the facility would work with the industrial contractors to develop and manage the most challenging project in electrical power requirements for any space programme to date.

The power management and distribution system for the ISS provides 160vdc through a series of switches that have integral microprocessors controlled by software linked to a computer network running throughout the ISS. A step-down transformer conditions the voltage to 120vdc forming a secondary power source to service appropriate loads. These converters also isolate the secondary system from the primary for uniform power quality throughout the station.

Early work on space station power concepts supported contractor studies at Lewis from the dawn of the Space Age and matured to real-world power provision with the Mir Cooperative Solar Array (MCSA) which was delivered to the Russian space station on STS-74 during

LEFT The Plum Brook facility reported to Glenn for development of a research reactor supporting interest in nuclear rocket propulsion. (NASA-GRC)

a Shuttle/Mir docking in November 1995. The Russians recorded data during the mission and sent it to Lewis for comparative analysis against computer predictions, data which helped validate the evolving design configuration for the ISS power system.

Lewis developed the System Power Analysis for Capability Evolution (SPACE) computer code used as a tool to predict the maximum power levels that could be achieved on the ISS for sustained continuity during the continual switch between the solar arrays and the nickel-hydrogen rechargeable batteries, covering eclipse periods and the day/night cycle when the station was in darkness.

As designed by the Lewis Center, the ISS has eight solar arrays, each 112ft (34.1m) long and 12ft (3.6m) wide, a total of more than 250,000 photovoltaic cells. The complete power system, including the Russian arrays, delivers 110kW of total power, of which about 46kW is available for scientific experiments and associated equipment. Lewis was also responsible for the seven-panel radiator system for cooling the systems and components on board. It was tested at the Plum Brook facility in Sandusky, Ohio. This is the world's largest space simulation chamber (the Thermal-Vacuum Test Chamber), with a height of 122ft (37.2m) and a diameter of 100ft (30m).

Plum Brook was set up specifically to carry out research into nuclear power and propulsion

for aircraft, utilising a 60MW reactor. Some 18 locations were surveyed prior to the NACA leasing 500 acres (202.3ha) from the US Army in March 1956. Named after a small stream running through the property, it had been used as a TNT manufacturing facility and ground was broken on 26 September 1956 for the Plum Brook Research Reactor Facility. The reactor became operational on 14 June 1961 but work shifted to propulsion for rocket stages and power production on large orbiting facilities.

BELOW The Kiwi-A prime nuclear reactor which supported research at Glenn on development of a NERVA rocket motor operating on hydrogen fuel but without the need for an oxidiser. (NASA-GRC)

RIGHT Glenn has a collection of space memorabilia, including the Apollo Command Module which returned the last Skylab crew to Earth in 1974.
(NASA-GRC)

On 1 March 1999 Lewis got another name change, to the John H. Glenn Research Center, commemorating the achievements of the former test pilot and NASA's first astronaut to orbit the Earth. Today, the facility is a diverse and energetic centre for a wide range of research activities which includes some unique capabilities at its Plum Brook outstation.

Not least is the Zero Gravity Facility (ZGF) with a height of 510ft (160m) below ground encapsulating a steel vacuum chamber with a diameter of 20ft (6.1m) which has a drop height of 460ft (140.2m) providing 5.18 seconds of zero gravity and up to 32g deceleration. Test equipment is decelerated at the base by a bed of expanded Styrofoam beads. It was built in 1966 to support the Centaur development programme by studying the weightless behaviour of fluids and was awarded National Historic Landmark status in 1985.

Sadly, the NASA Glenn Visitor Center closed in September 2009 but most of the exhibits were transferred to the Great Lakes Science Center which merged these with new displays. In an integrated section which uses the same name, it now attracts more than 300,000 visitors a year compared with 60,000 when it was at the NASA facility.

RIGHT The collection of space exhibits from Glenn has now moved to the Great Lakes Science Center where it is receiving many more visitors than it did when hosted by the NASA facility.
(NASA-GRC)

Wallops Flight Facility

Located 75 miles (120km) north-east of Langley Research Center (LRC), when NASA began it had been known as the Pilotless Aircraft Research Station from the date it was set up in 1945. Managed by Langley, it boasted 80 people and had an annual operating budget of $3.5million. While 90% of its work was assessed as space-related, this was the experimental test range for high-altitude sounding rockets and for obtaining telemetered data on conditions in the upper atmosphere. It was the proving ground for many ideas originating at the LRC which would directly lead to the NACA's studies into manned space flight, inspired by the Air Force MISS programme described earlier.

The origin of this facility harks back to the days when rocketry was emerging amid an almost complete absence of technical information about high-speed missile research and Langley Memorial Aeronautical Laboratory formed a "Special Flying Weapons Team" which met on 9 December 1944 to develop a research programme to fill the yawning gaps between what was known from the engineering world and what was demanded by the Army.

The NACA approved such a facility on 25 January 1945 and Congress appropriated funds

ABOVE **When the NACA began to build up facilities at Wallops Island the place was a sandy beach with no on-site comforts but it was the first launch location specifically set up in the United States for rocketry.** *(NASA-Wallops)*

on 25 April for development at a site which was then under consideration by the Navy as its primary guided missile station. Wallops Island was selected, a narrow 6 mile (9.7km) strip of land off the eastern shore of Virginia. It had been named after John Wallop, the recipient of a Crown patent from King Charles II of England in 1672, the entire island being acquired by the Wallops Island Association in 1889. On 11 May

BELOW **A Terrier-Improved Orion suborbital sounding rocket is launched from Wallops carrying 17 experiments from several universities and students on 23 June 2011.** *(NASA-Wallops)*

LEFT A Minotaur V rocket from Orbital Sciences on the pad at Wallops prior to the launch of the LADEE satellite on 7 September 2013, which was destined to orbit the Moon and obtain scientific data about its structure and composition. *(Orbital-ATK)*

1945, 1,000 acres (404.7ha) was leased to the government and despite the Navy wanting to purchase the complete island the NACA received approval on 18 September to possess 84.87 acres (34.3ha) although construction of facilities was not completed before 1947.

Established on 7 May 1945 as the Auxiliary Flight Research Station, a unit of Langley, there was sufficient construction progress for the site to host its first firing, eight 3.25in (83mm) rockets, on 27 June 1945. Tests got under way with a missile called Tiamat on 4 July and flights with a dummy RM-1 (Research Missile-1) began on 17 October 1945, instrumented launches getting under way the following day, also accompanied by RM-2 on 18 October 1945. Research into the aerodynamic control of high-speed rockets started in May 1946 with the RM-5 and tests into the supersonic regime got under way with the RM-6 to RM-10 projects in 1947.

On 10 June 1946 the facility had changed its name to the Pilotless Aircraft Research Division (PARD) with the site now called the Pilotless Aircraft Research Station, or simply "Wallops". Direct support of aviation research began on 25 April 1947 with a rocket-propelled model of the Republic XF-91 Thunderceptor and within months almost every new Air Force and Navy combat jet was being tested in model form at Wallops. To conduct research beyond the operating envelope of extant aircraft prototypes, a rocket called Deacon was developed and used to investigate high-speed aerodynamics.

By mid-1952, the NACA had formalised the corridors of the flight envelopes which should receive primary attention as being the 12–50 mile (20-80km) altitude band and from Mach 10

LEFT Many Scout rockets were launched from Wallops, which became a vital component in the inventory of US launch facilities. *(NASA-Wallops)*

to Earth escape velocity in speed. All NACA laboratories were so informed and the front end of this push came to be a significant part of an agency-wide drive to prepare the nation for hypersonic flight through the atmosphere and significant steps forward into space. At no place was that more self-evident than the Langley-Wallops tie-up. But there was the advantage of an agency-wide drive which created a broad consensus and support infrastructure.

There is no better example of that than the work at Ames on blunt-body research toward survivable re-entry devices for ballistic missiles carrying complex warheads. As velocities increased the kinetic energy on an incoming projectile increased to levels where the heat incurred was beyond the melting point of any known material. Research into how to survive a fiery re-entry from near-orbital velocities would unlock opportunities for intercontinental ballistic missiles, their warheads diving back down through the atmosphere toward targets in a different hemisphere.

PARD took up flight tests using multi-stage, solid-propellant rockets studying heat transfer from the kinetic energy experienced on re-entry, the capability to survive such extremes also opening the door to human space travel, for it would be in much larger versions of ballistic warheads that people would first fly into space.

ABOVE An Antares rocket carrying a Cygnus cargo module for the International Space Station begins its journey to the launch pad.
(Orbital-ATK)

LEFT Antares ready to fly its Cygnus cargo module to the ISS, just one of the commercial contracts secured to link government to private enterprise.
(Orbital-ATK)

The first launch of a three-stage vehicle for these tests was conducted on 29 April 1954 and on 24 August 1956 PARD tested a five-stage rocket, achieving a speed of Mach 15. This work fed directly back through Langley to the other NACA field centres, the results informing Air Force plans for high-performance vehicles and re-entry warheads for ballistic missiles, thoroughly encouraging plans for space exploration.

When NASA formed on 1 October 1958, PARD kept its name as 14 employees were transferred to the group soon to be known as the Space Task Group organising the Mercury programme, inherited from the Air Force's Man-In-Space-Soonest (MISS) programme, which the NACA had been supporting. As the work continued, additional resources grew with the priorities of the nascent space programme

and on 4 October 1959 the facility began flight tests with the Little Joe solid propellant boost rocket for testing the Mercury capsule during its development.

Pushing further into the exotic realm of re-entry physics, on 3 March 1959 Wallops flew its first six-stage, solid propellant rocket to a maximum Mach 26 and within a year had begun organising the first Scout flights of this low-cost small satellite launcher. During the 1960s alone Wallops launched more than 300 experiments every year studying the atmosphere and the space environment. In all, the facility had five launch locations for sounding rockets, experimental flights and small satellite launches, supported by an elaborate and comprehensive array of communications and telemetry facilities.

By the end of the 1960s the total area had grown to 6,613 acres (2,767ha) with employment having more than doubled to over 400 people but change came to Wallops as it had to every other NASA facility after the Apollo programme. Over the next several decades the broad-based activity at Wallops embraced research for the National Oceanic and Atmospheric Administration (NOAA) as well as activities on behalf of other countries. Support for sounding rocket flights from other places in the world exercises the tracking and data collection facilities at this location and the site now hosts around 1,000 employees working a wide range of projects with research at the very heart of everything they do there.

Armstrong Flight Research Center

When NASA was formed in 1958 this facility was known as the High-Speed Flight Research Station and ran with 300 employees at an annual budget of $16.5million. Its work was primarily at the cutting edge of flight and flying, which was its mandate. It was here that the celebrated X-series research aircraft were evaluated in the air and from where many important milestones in aviation were laid. Under the existing definition at the time, 42% of its work was related to space research but this was almost exclusively in high-speed, high-altitude flight using a variety of experimental aircraft and in developing flight control systems and related equipment to push the boundaries of the theoretical envelope.

The story of this place began in September 1946 when 13 engineers and scientists were sent from the NACA's Langley Aeronautical Laboratory to the Army Air Force's flight test range at Muroc, California. They were sent there on temporary assignment and as the Muroc Flight Test Unit (MFTU) they presided over the attempt on the "sound barrier" with the Bell XS-1, which made its first powered flight on 9 December 1946. The NACA had been deeply involved with the search for a way to push aircraft beyond the speed of sound (Mach 1) and their engineers were central to this Air Force programme.

Flight tests with the XS-1 preceded the first successful flight through Mach 1 on 14 October 1947 with Capt Charles E. Yeager at the controls and, by the end of the year, the MFTU was given official status, expanding to 60 personnel by the end of 1948. The following year the NACA group was established as the NACA High Speed Flight Research Station (HSFRS), a division of Langley Aeronautical Laboratory. Muroc was renamed Edwards Air Force Base on 27 January 1950 in honour of test pilot Capt Glen W. Edwards, who had been killed in the crash of a Northrop YB-49 flying wing bomber prototype on 5 June 1948.

The Army had been at Muroc since 1933 when Col Henry H. Arnold, commanding officer of March Field, wanted to use this desolate location for bombing and gunnery practice, seeing the vast and isolated expanse of the Mojave Desert around Rogers Dry Lake as an ideal site. When Arnold became chief of the Army Air Forces and learned about British developments with the jet engine he assigned tests of a US jet fighter, the XP-59A Airacomet, to Muroc rather than the traditional centre for Air Force research and development, Wright Field, Dayton, Ohio, which was in a heavily residential area.

Gradually, Muroc became more important than Wright Field and developed the North Base as its advanced flight test centre distinctly separate from the South Base, which was reserved for training purposes. The advent of the jet age and the pressures of the Cold War deemed places such as Muroc-Edwards vital assets in development of secret projects. To the north of the site, alongside the Sierra Nevada Mountains, the China Lake Naval Weapons Center was set up while in a different place Plant 42 not far from Palmdale became a production and development facility for military aircraft. Soon, Area 51 would add to the mystique of this assemblage of high-performance, classified projects.

On 1 July 1954 the NACA facility at Edwards was renamed the High Speed Flight Station (HSFS) and separated from Langley to become an autonomous NACA facility, endorsed when 250 NACA employees moved from shared accommodation to their own research facilities

BELOW Edwards Air Force Base and the adjacent Rogers Dry Lake have been home to the NACA and its success for more than 70 years, pioneering landmark achievements with flight and flying and supporting many research projects aiming to send pilots beyond the atmosphere. *(NASA-AFRC)*

midway between North Base and South Base. Four years later, after the NACA transitioned to become NASA, on 27 September 1959 the HSFS became the Flight Research Center (FRS) and experimental flight operations began to take on a new level of research and experimental activity involving increasing numbers of scientists and engineers.

Initially, the biggest high-performance flight test programme at the FRC was the North American Aviation X-15, a rocket-propelled hypersonic research vehicle capable of exiting the atmosphere for a brief dip into true space before returning to a conventional landing. But like many of the experimental X-series research aircraft before it, the X-15 would be carried

LEFT Ground crew pose for a rare shot of the unsung heroes who routinely sent aircraft like the XS-1 into harm's way, the pilot's safety dependent on the thoroughness of their work. This shot of the XS-1 was taken in September 1949. *(NASA-AFRC)*

RIGHT Hangar Queens in 1953 – but not for long! From left to right clockwise: three Douglas D-558-2 rocket-powered research aircraft; Douglas D-558-1 (minus nose); Boeing B-47 Stratojet; wing of a Republic YF-84A. To the rear are a tailless Northrop X-4 Bantam and North American F-51 (re-designated from its original wartime prefix as P-51). *(NASA-AFRC)*

into the air by a mothership, in this case one of two converted B-52s, a B-52A and a B-52B, re-designated with the prefix N. The first X-15 glide flight took place on 8 June 1959 with the first powered flight on 17 September.

Three X-15s were built accumulating 199 flights, the last on 24 October 1968, in which time it evaluated the aero-thermal environment at speeds in excess of Mach 5 and to altitudes above 354,000ft (107,900m) or 67 miles. Over time, eight pilots made 12 flights above 50 miles (80km) which at the time was set as the somewhat arbitrary line of demarcation between the atmosphere and the vacuum of space. When the *Fédération Aéronautique Internationale* decided that this line should be placed at a rounded 100km (62.1 miles) it still resulted in NASA test pilot Joe Walker being credited with making two space flights. One of

the three X-15s was destroyed on 9 November 1962, killing pilot Jack McKay.

The measurements of the piloted flight environment into space made far above the altitude where a conventional aircraft could achieve sustained lifting flight would pave the way later for development of the NASA Shuttle and this research was vital to what was to come with the ability of skilled pilots to achieve controlled descent through the atmosphere. While the X-15 had been under way since 1954 in a joint venture with the Air Force, transitioning to a NASA-run research programme, other activity at the Flight Research Center was directly stimulated by research at other field centres.

One such was the manned flights with the M2-F1 lifting body which began in 1963, initial flight tests with the heavier M2-F1 beginning on 12 July 1966, and with the HL-10 lifting body

ABOVE Four decades on, another cluster significantly advanced beyond the flying types that populated the 1950s. Left to right clockwise: Rockwell-MBB X-31, McDonnell Douglas F-15 ACTIVE, Lockheed SR-71, General Dynamics F-106, Lockheed F-16XL, NASA X-38. In centre: McDonnell Douglas X-36 and radio controlled model. *(NASA-AFRC)*

ABOVE One of the several control rooms at the centre, on this occasion personnel monitoring flight activity with the Grumman X-29. *(NASA-AFRC)*

RIGHT Technician Jeff Greulich adjusts the gear of a test pilot prior to flight trials with a new helmet design, one of the many activities which usually escapes public attention but which inevitably brings enormous benefit to those whose professional life is flight and flying. *(NASA-AFRC)*

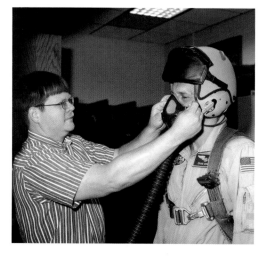

on 22 December 1966. The Air Force X-42A joined the group with flight tests beginning 17 April 1969. The lifting-body concept originated at the Ames Research Center (which see) in the search for a reusable space vehicle fully controllable and capable of making a land landing rather than the splashdown landing of a ballistic spacecraft such as Mercury, Gemini and Apollo. Lifting-body flight tests continued until 26 November 1975 when a total of 144 flights had been made, some in excess of Mach 1.5 and many at high altitude.

Many other research programmes were carried out at this facility, including flight tests with the two North American XB-70A intercontinental bomber prototypes. Originating in the mid-1950s as a strategic bomber capable of cruising at Mach 3 and flying at altitudes up to 77,350ft (23,600m), the XB-70 was powered by six YJ93 engines, each delivering a dry thrust of 19,900lb (88.5kN) and possessing a maximum range of 4,288 miles (6,900km) carrying the highest yield thermonuclear weapons available. The aircraft had a wingspan of 105ft (32m) and a length of 185.9ft (56.6m) and weighed 550,000lb (250,000kg). The programme was cancelled in March 1961 but retained as a research and development project. NASA supported 23 flights of the #1 aircraft at the FRC between 25 April 1967 and 4 February 1969. These came on the back of 49 flights by the #1 aircraft and 46 by the #2 before it was lost in a flying accident, ironically during a public relations photo-shoot on 8 June 1966.

Arguably the most advanced conventional aircraft evaluated by NASA was the Lockheed YF-12, capable of sustained flight at speeds just above Mach 3 and a service ceiling of 85,000ft (25,900m). The type evolved from the A12, conceived in the late 1950s as a fast reconnaissance aircraft possessing low-observables (stealth) characteristics and therefore less easy to detect by enemy radar. The surge in high-Mach number military aircraft

LEFT The Lunar Landing Research Vehicle had a vertically mounted jet engine to support five-sixths of its mass and provide a realistic simulation of flying the terminal descent phase of a Moon landing under throttle. *(NASA-AFRC)*

resulted from major research programmes within industry, as well as fundamental flight data at supersonic speeds obtained by NASA at its Flight Research Center. The A-12 was, essentially, an unarmed successor to the Lockheed U-2 spyplane but the Air Force adopted the type as an advanced interceptor designated the YF-12A, the first flown at Groom Lake (Area 51) on 7 August 1963.

The YF-12A was never introduced as an operational interceptor but NASA took two of the three built plus a converted SR-71, which had itself evolved from the A-12 and was re-designated YF-12C. NASA's involvement with the programme extended from 1969 to 1979, with 297 flights in that period totalling about 450 flying hours, of which 37 hours was spent at or above Mach 3. Pilots for the programme were Fitzhugh Fulton and Donald Mallick, with flight-test engineers Victor Horton and Ray Young. The last flight of a YF-12A was on 7 November 1979 when one was flown to the USAF Museum, Wright-Patterson AFB, Dayton, Ohio, by Col J. Sullivan and Col R. Uppstrom.

The combined research of the Lifting Bodies tested at the Flight Research Center and the aerodynamic and thermodynamic studies undertaken at Langley and Ames converged toward a design in the mid-1960s to devise a way of reusing rockets employed to carry satellites and spacecraft into orbit. There were two ways of achieving reusability: bring back the booster rockets and their engines on parachutes or inflatable parasail-like devices;

or design a spaceplane that could shuttle back and forth, launched into space vertically like a conventional, expendable rocket, returning through the atmosphere using wings for controlled descent and a conventional landing on a runway.

By the late 1960s, while Apollo astronauts were getting ready to walk on the Moon, the Space Shuttle emerged as NASA's next major human space flight initiative and this required new technologies and untested design initiatives. As noted earlier, President Nixon approved development of the Shuttle in January 1972. Within a few months the configuration had been decided: the Shuttle would be powered by a reusable Space Shuttle Main Engine (SSME), three of which would be attached to the aft section of a double-delta

ABOVE Maj Jerauld Gentry pauses alongside the HL-10 lifting-body, a research programme originating in the 1960s which provided valuable air data for the design of reusable winged space vehicles such as the Shuttle and the X-37B. *(NASA-AFRC)*

LEFT The fastest aircraft in the world – the North American X-15 – which conducted 199 flights, the last in October 1968, lifted into the air on the starboard underwing pylon of a B-52 carrier-plane. *(NASA-AFRC)*

winged Orbiter about the size of a small airliner, such as the Boeing 737-400; an External propellant Tank (ET) to provide propellant for the SSMEs and to which the Orbiter would be attached; and two Solid Rocket Boosters (SRBs), one either side of the ET.

The Orbiter would be flight tested in advance of its first launch and for that it was decided to fly it off the back of a specially converted Boeing 747-100 in a series of tests from the Flight Research Center. In any event, Edwards AFB would be used for the return of the Shuttle from space for at least the first several orbital missions, taking advantage of the wide open areas for possible errors in terminal guidance and landing. The use of this facility to evaluate the performance of the Orbiter in simulation of the terminal stage of the flight path would verify the design of the vehicle and give the pilots a feel for its handling. The converted Boeing 747-100 would be used routinely to fly the Orbiters back to the Kennedy Space Center so the effectiveness of handling these two air vehicles would be valuable.

Bombers had been used as carrier-planes for experimental aircraft since the X-1 had been lifted into the air underneath a converted B-29 and the X-15 under the starboard wing of an NB-52. But this would be very different, the Shuttle Orbiter being carried on top of the carrier-plane's fuselage using fixtures designed to attach it to the External Tank on the pad and carried all the way to orbit. To simulate the Shuttle's flight path after re-entering Earth's

atmosphere, the carrier-plane would release the Orbiter and nose-dive away and to one side, leaving the unpowered vehicle to start its own glide down toward the runway at Edwards.

The Air Launched Test (ALT) programme began on 18 February 1977 when OV-101 *Enterprise* was carried into the air in what was the first of five "captive-inactive" flights evaluating the handling characteristics of the mated configuration at increasingly higher altitude up to 30,000ft (9,100m) on the final such flight on 2 March. Next came three "captive-active" tests between 18 June and 26 July in which the flight deck of *Enterprise* was powered up and occupied by a pair of astronauts, either Fred Haise and Charles Fullerton or Joseph Engle and Richard Truly. Haise had been on the near-fatal Apollo 13 and Engle had been an X-15 pilot before being bumped from Apollo 17 so that a geologist – Harrison Schmitt – could work on the Moon.

The third and final phase of the ALT programme involved five free-flights from off the top of the carrier-plane, the first occurring on 12 August 1977, a historic day when the Shuttle made its first flight, albeit from a speed of 311mph (500kph) and a drop-altitude of 24,100ft (7,346m) piloted by Haise and Fullerton. Two more free-flights followed, on 13 September and 23 September, from about the same altitude.

All flights on the carrier-plane had been made with a cone giving the rear of *Enterprise* a streamlined appearance, reducing buffeting on the Boeing 747-100 but also reducing drag on the Orbiter when it flew free. The last two free-flight tests were made with the tail cone off and three simulated engines in the rear, giving it a blunt aft end and considerably increasing the drag after separation from the carrier-plane. The effect of this was to more realistically simulate the terminal descent of an Orbiter returning from space and significantly increasing the inclination of the glideslope so that *Enterprise* fell to Earth at a descent rate of more than 9,000ft (2,740m) a minute!

The first free-flight with the tail cone off occurred on 12 October followed by the second and last such demonstration on 26 October. All five drop tests had been successful, considerable data was obtained and the Shuttle

BELOW The Space Shuttle Orbiter Enterprise separates from the top of the converted Boeing 747 in Air Launched Test flights conducted in 1977. *(NASA-AFRC)*

moved a step closer to orbital flight courtesy of the Flight Research Center. *Enterprise* was too heavy to be converted for space flight and lacked many of the systems operational Orbiters would acquire as they were manufactured. This pioneer would never make it into space.

It was destined to carry out vibration tests for the Marshall Space Flight Center where it was delivered in March 1978 and from there, in April 1979, it was flown by carrier-plane to the Kennedy Space Center for fit-checks with the first External Tank built for the maiden flight and two inert SRBs. It was in this configuration that the world got to see a Shuttle roll out from the Vehicle Assembly Building to LC-39A on 23 July 1979, the pad which ten years earlier had launched Apollo 11 to the Moon. After that it was flown to Vandenberg Air Force Base (VAFB) for pad checkout tests where NASA planned to launch Shuttles to polar orbit. They never did.

By this time the Flight Research Center had been renamed the Dryden Flight Research Center, in honour of Hugh L. Dryden, the last man to head the NACA before it became NASA. The inauguration ceremony took place on 26 March 1976. But this decade saw a transitional change in NASA's aeronautical activity as flight test and evaluation at DFRC settled into a new routine, but nonetheless just as important and challenging a set of experimental test programmes as it always had been.

For the duration of the Shuttle programme, after the ALT flights DFRC supported orbital operations between 1981 and 2011 and of

the 133 Orbiter landings, 54 (40.6%) were at Edwards AFB of which seven were at night. The first landing had been that of *Columbia* (STS-1) on 14 April 1981, with the last conducted by Discovery (STS-128) when it touched down on 11 September 2009. For the last seven Shuttle flights, DFRC was on standby as an emergency landing site.

On 1 March 2014, the DFRC was renamed the Armstrong Flight Research Center in honour of the first man to walk on the Moon. Today, the Armstrong Center continues to support new generations of high-performance aircraft and is considered by NASA as its primary aircraft development centre, collectively picking up on a lot of basic research from other facilities which have underpinned the modern aviation environment.

ABOVE Inelegantly named the Mate/ De-Mate Facility, this structure raised the Shuttle for installation on top of the converted Boeing 747, employed to return Orbiters to Cape Canaveral, should they land at Edwards from orbit, also used as seen here after STS-126 to inspect the main engines. *(NASA-AFRC)*

LEFT Exotic research programmes are not new to the Armstrong Flight Research Center, epitomised here by this Lockheed SR-71 equipped with a test rig to measure the aerodynamic flow from the plume of an aerospike engine. *(NASA-AFRC)*

ABOVE NASA Headquarters is reliant upon the US Congress for the funding to carry out its many science and research programmes, a balance between government proposals from the executive led by the President and the choices made by the legislature in both Houses. *(Via David Baker)*

BELOW NASA has an organisational structure which has evolved across many decades to facilitate a changing spectrum of programmes, this present configuration balancing human and robotic activity along with science, life science, technology and engineering. *(NASA)*

The NACA had been managed through a headquarters located in Washington, DC, since the organisation had been formed in 1915, with John F. Victory as clerk. In 1958 there were 170 employees. This office ran the general administration of the NACA and left the separate field offices to manage their own activities under the remit of the various committees. In general, relations between the HQ and these dispersed locations was good, a condition which would not always prevail under the greatly expanded structure of NASA. HQ also served to connect the two liaison offices at Dayton, Ohio, the location of the Wright-Patterson Air Force Base, and Los Angeles, California, operating as the Western Coordination Office.

When NASA formed, HQ did not have the technical or scientific strength to get involved in the day-to-day running of programmes and projects. Only 30 of the staff were professional aeronautical people but they did maintain a very close working relationship and the Top Three

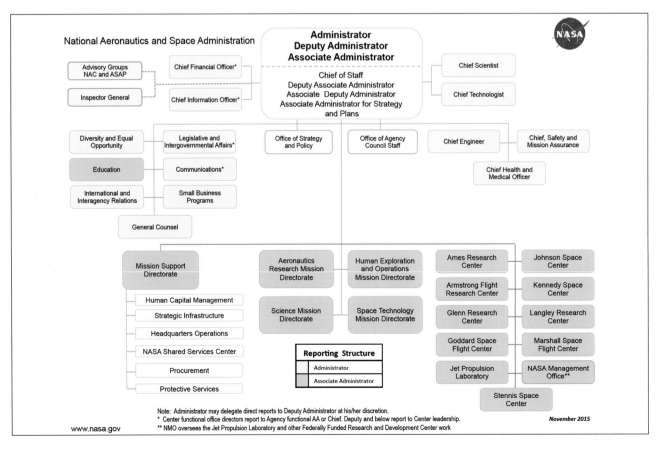

at the old NACA had been able to operate the place as a large staff office managing the work carried out at the field centres. But the desire for independence, which would come to plague some periods in NASA history, began to appear as field centres vehemently opposed some impositions from headquarters.

During the NACA years, headquarters was located in DC because it was funded through appropriations and authorisation from Congress and it was in close proximity to the then Bureau of the Budget, which was equally important. Later, especially in the early days of NASA, it became the connecting node between the legislature and the field centres, accountable to the executive branch of government but shaped largely by the legislature on Capitol Hill. Frequently, it would find itself embroiled in the different perspectives about direction posited by the White House and challenges by Congress.

The NACA had been resident at 1724 F Street NW until June 1954 when it moved to the Wilkins Building at 1512 H Street NW and it was there when Sputnik was launched to begin a new chapter in space activity. From 29 July 1958 when the Space Act was signed the planning group in charge of setting up NASA sought additional building space and in September Administrator Glennan occupied the newly acquired Dolley Madison House at 1520 H Street NW. This building had been built in 1830 by Benjamin Tayloe and had been occupied by the wife of President James Madison from 1837 to 1849.

At the turn of the century it was known as the Cosmos Club and had provided temporary accommodation for the Wright Brothers, becoming the official home of NASA from 1 October 1958. Between July and October 1961 a part of HQ moved to the newly completed Federal Office Building (FOB) No 6 which it shared with the Department of Health, Education & Welfare but on

8 November 1963 it began to share occupancy of FOB No 10B and personnel who since 1963 had occupied the Universal Building North at 1875 Connecticut Avenue NW moved in October and November 1965 to the Reporters Building at 300 7th Street SW near the two FOBs.

The Procurement Division moved from the Universal Building to 10B in April 1966 and in April 1968 the Apollo Program Office began a series of moves from 10B to the L'Enfant Plaza North Building, a transfer which brought its own complexities and headaches as this writer remembers. Paradoxically, the budget was winding down and there were several administrative shifts which confounded efforts to maintain a smooth and trouble-free migration!

Much of the work of HQ was related to the budget and that aspect of NASA history can be found in Section 4 but the early challenges grew from the need to form an integrated agency from the old NACA and the Naval Research Laboratory and manage the transfer of the Jet Propulsion Laboratory and the Development Operations Division of the Army Ballistic Missile Agency. The initial organisational charts defined the Top

BELOW NASA managers regularly address major space issues at hearings before special Congressional committees and to Congresspersons managing the authorisation and appropriations of the agency awarded each financial year. These financial years start on 1 October and are known by the year following, fiscal year 2019 beginning on 1 October 2018. *(NASA)*

RIGHT Revived by President Donald Trump, with Mike Pence as its chair, the National Space Council has rejuvenated discussion about the balance between military and civilian programmes, and about the way in which commercial space programmes can save the taxpayer money, stimulate industry and increase efficiency. *(NASA)*

ABOVE NASA Administrator James Bridenstine addresses some of his staff at NASA Headquarters, the 13th person to hold the top job at the agency, not including temporary acting positions between formal appointments. *(NASA)*

Three, as described on the previous pages and more comprehensively on page 26. But in 1961 HQ assumed greater responsibility as the central coordinating authority when President Kennedy made his Moon goal speech on 25 May.

The sudden demand for growth prompted the formation of the four principal offices which would see it through its first decade and beyond: Manned Space Flight, Space Sciences, Applications, and Advanced Research and Technology. The need for a much bigger HQ and an expanded staffing level came upon NASA unexpectedly – it had all happened within 43 days, from the flight of Yuri Gagarin to the announcement of the Moon goal via a

memorandum to Vice President Johnson seeking ways to eclipse Soviet successes. Only at the very top had management known what was coming and many personnel were caught out.

When NASA was formed in October 1958, HQ employed only 180 people, about 2% of the total agency of fewer than 8,000 people. The permanent staff included nine people with excepted positions and 37 research scientists and engineers. By December 1961 the HQ staffing level increased to 922 and a year later had reached 1,641. The total number of people working for NASA peaked at 33,722 in December 1966, when HQ had 2,152 permanent positions, more than 6% of the agency total. As the budget began to decline, HQ staffing dropped to 2,077 by 30 June 1968.

As always, NASA HQ is at the centre of decisions about the agency's plans and programmes, liaising with and reporting to Congress and responsive to the direction defined by the Office of the President. It is led by the Administrator, currently Jim Bridenstine, with a Deputy Administrator and an Associate Administrator, Deputy Associate Administrator, Chief of Staff and the Associate Administrator for Strategy and Plans operating out of the Office of the Administrator.

Two advisory groups, the NASA Advisory Council (NAC) and the Aerospace Advisory Safety Panel (ASAP) report directly to the Office of the Administrator. The NAC meets regularly to review and report on the activities and direction of agency activity and various reports are released at frequent intervals, identifying key outliers in various aeronautical programmes as well as those directly connected to space activity. It has committees focusing on aeronautics, human exploration and operations, science, technology and engineering, and on STEM education.

Task Force reports on STEM (science, technology, engineering and maths) subjects in education are also under the NAC, as are the strategic plans and proposed budgets. The NAC sets the general overview through analysis and observation, its members having an extensive background in respective specialities.

The ASAP was formed in 1968 after the Apollo fire and monitors the agency's safety performance while inspecting various programmes to oversee the levels of safety and

BELOW One of the demands on an Administrator's job is not only to hold the agency together confidently and with tolerance for disparate views but also to report the progress of the agency to the government and to the public, demands that all too frequently conflict with management of NASA itself. *(NASA)*

quality assurance built in to NASA's projects, frequently reporting on critical programmes such as the Space Launch System and various human space flight initiatives including those of the commercial sector with which NASA has contracted services. It frequently expresses concerns and points to critical issues regarding safety, while making observations regarding the general operational philosophy at NASA as it relates to risk mitigation.

Headquarters has mission directorates in aeronautical research, human exploration and operations, science, and space technology. It is through these that NASA's various projects and programmes are managed, with field centres receiving a distribution of work according to their specialities and skills, the facilities they have at their disposal and the connections to industrial contractors.

NASA has a public voice through its information systems, its web-based connections to various projects and programmes and through NASA TV, which is available free to air around the world. It has a History Office which contains materials relevant to HQ and the general activities of NASA, with resources available to bona fide researchers, writers and reporters. It also commissions a series of books and monographs on various projects and has made extensive efforts to place as much of these online as possible, many of them as free downloads.

NASA HQ is responsible for reporting to

ABOVE Former astronaut and NASA Administrator Charlie Bolden managed the agency during the difficult transition from the Shuttle era to the new commitment to renewed deep-space exploration that will carry astronauts back to the Moon and on to Mars, here addressing personnel at the Kennedy Space Center with KSC Director, and former astronaut, Robert Cabana to his left. *(NASA-KSC)*

Congressional committees each year on requests from the President for funding and programme support, defined by the White House and the executive, with the legislature responsible for managing the balance between what it sees as in the best interests of the nation and the requests from the President. Because NASA is a government agency, its direction of travel is a delicate balance between the White House and Congress and there have been numerous occasions where the requests of the President have been rejected or overturned by one or both Houses of Congress.

BELOW Conducted by Emil de Cou, the US National Symphony Orchestra played at the John F. Kennedy Center for the Performing Arts, Washington, DC, on 1 June 2018 in celebration of NASA's 60th anniversary. *(NASA)*

RIGHT Located on Oak Grove Drive, Pasadena, California and close to the Hollywood Bowl where open air concerts frequently send musical tones over cool evenings, JPL is unique among all the sites NASA does business – idyllic and aesthetic, a place of learning and discovery at the frontiers of our solar system. *(NASA-JPL)*

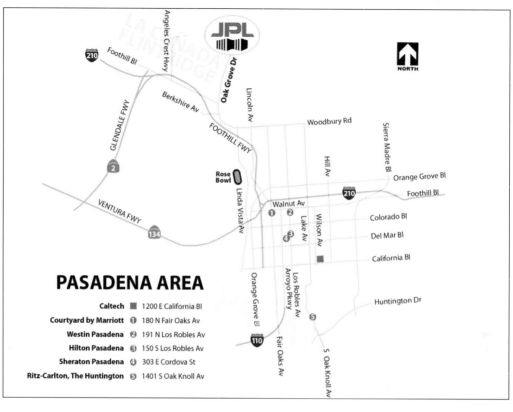

PASADENA AREA

Caltech	■	1200 E California Bl
Courtyard by Marriott	❶	180 N Fair Oaks Av
Westin Pasadena	❷	191 N Los Robles Av
Hilton Pasadena	❸	150 S Los Robles Av
Sheraton Pasadena	❹	303 E Cordova St
Ritz-Carlton, The Huntington	❺	1401 S Oak Knoll Av

BELOW Nestled within scenic country north of Los Angeles, the Jet Propulsion Laboratory has been the focus for planetary scientists in the United States and from around the world as international teams work together on flights throughout the solar system and beyond. *(NASA-JPL)*

Certain sites across the United States are landmarks in the story of America's race for space, a journey which began in the 1930s when mad-cap inventors, enthusiasts and the first rocket engineers began to experiment with propellants, only dreaming of a time in which they could travel to the Moon and the planets. A key location for those activities was the Jet Propulsion Laboratory (JPL), a place that would become synonymous with America's first satellite and the robotic exploration of the Moon and the solar system, truly realising the dreams of its founders. Yet the Jet Propulsion Laboratory is not a NASA centre.

The origin of JPL dates back to 1936 when experiments began at California Institute of Technology (Caltech) in the Guggenheim Aeronautical Laboratory, known most commonly as GALCIT. It aspired to become a national leader as a school of aeronautics and, under Dr Theodore von Kármán, research on jet propulsion and rocketry was encouraged. Early research resulted in a contract from the

Army, awarded on 25 June 1940, for solid and liquid propellant rockets for aircraft, initially for boosting take-off speed on short runways.

This work gave GALCIT a lead in solid propellant rockets and in hypergolic motors using red-fuming-nitric-acid-aniline motors. From 1942 production was given to the Aerojet-Engineering Corporation, later renamed Aerojet-General. Liquid propellant rocket motors were known to have greater potential for long-range flight, given the greater efficiency and working potential compared to solid propellant rockets. When GALCIT became aware of the German liquid propellant V-2 rockets they conducted a detailed study of long-range missiles and lobbied the Army for funds to support development of experimental rockets of this type.

On 22 June 1944 GALCIT received a contract for that and for appropriate launching equipment which led to a focus on this class of rocket to an extent that a name change was mooted, with Rocket Propulsion Laboratory being proposed. Given the "Buck Rogers" association with comic-books that rocketry invoked in the public and academic mind, and the knowledge that by this time jet fighters were a reality, the title Jet Propulsion Laboratory was chosen instead and that became official from 1 November 1944.

The contract resulted in development of the Corporal, America's first tactical missile, and the WAC Corporal, a sounding rocket for carrying instruments to the upper atmosphere and the fringe of space. The name derives from an aberration in the naming sequence which was then becoming Army practice, calling successive rockets getting larger as the designs progressed by names for enlisted ranks (Recruit, Private, Sergeant, and Corporal). But

BELOW President John F. Kennedy receives a model of the Mariner 2 spacecraft, gifted by Dr William Pickering (to whom he is facing), head of JPL, after the successful fly-by of Venus in 1962. Between the two against the window is NASA Administrator James Webb. (NASA-JPL)

the smaller-sized sounding rocket regressed from that trend toward bigger and more potent designs so the prefix WAC was chosen as an acronym for Women's Army Corps, indicating no further progression!

The military Corporal was the Army's first tactical nuclear battlefield missile and had a range of 86 miles (139km) and was sold to the British in 1954 and deployed by US Army units from 1955. The smaller WAC Corporal made its first flight on 16 September 1945 and on 22 May 1946 reached a height of 50 miles (80km) which was at that time considered to be the line between air and space. Attached as an upper stage to a Bumper rocket, a slightly modified captured V-2, on 24 February 1949 this combination reached an altitude of 244 miles (393km), the highest altitude achieved by a man-made object to that date, after launch from White Sands. After tests moved to the Long Range Proving Ground, a Bumper-WAC conducted the first launch from there on 24 July 1950.

JPL set up extensive tracking and data networks and pioneered radio and inertial guidance systems as well as FM-FM telemetry, technologies incorporated into the Corporal, Sergeant and eventually the Jupiter IRBM. JPL supported von Braun's bid to launch an artificial satellite for the International Geophysical Year and while that job went to Vanguard, an ostensibly civilian project out of the Viking sounding rocket programme, all the conceptual definition of how it could be done using a

Jupiter-C rocket with JPL solid propellant rockets in a cluster on top, was ready and available. When the first attempt at getting a Vanguard satellite in orbit failed the von Braun/JPL team got the job – and did it successfully, placing the first US satellite in orbit on 31 January 1958.

The executive order moving JPL to NASA was signed on 3 December 1958, effective from the end of that month, although in effect NASA contracted with Caltech for the services of JPL and that is how it would remain. Clearly qualified to take on the role of planetary science, JPL struck out early with plans for sending probes to the Moon, Venus and Mars, and later into deep-space exploration of the outer solar system. JPL undertook management of the Ranger and Surveyor Moon probes, Mariner missions to the inner planets and the later Voyager planets to Jupiter, Saturn, Uranus and Neptune. It also had responsibility for development of the Deep Space Network (DSN) which would maintain communication with spacecraft out to the most distant reaches of the solar system.

The primary driving force behind the rise of JPL in the space programme was the leadership of Dr William "Bill" Pickering, born in New Zealand but with dual citizenship, who managed JPL between 1 October 1954 and 31 March 1976. One of the true "greats" of the first two decades of NASA, Pickering is a legend and was followed by another equally driven manager, Dr Bruce Murray, who held the post to 30 June 1982. With major achievements in its portfolio of successful planetary programmes, JPL became the lead centre for asteroid detection and mapping, 95% of known near-Earth objects being attributed to this facility.

An unsung achievement has been its support for getting women into the space programme, employing an all-female computer group to calculate trajectories during the Ranger and Mariner missions of the early 1960s. But support for women engineers originated back in the 1940s and when computers began to take hold in the 1950s these women were hired out to schools and universities to teach staff operating procedures. Key in hiring women were Macie Roberts and Helen Ling, sustaining a tradition that really took hold after Barbara Paulson played a major role in the launch of

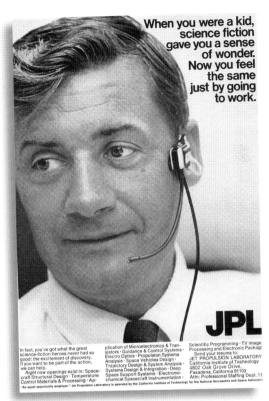

LEFT The outstanding success of lunar and planetary missions gifted advertisers with an easy "sell" when it came to recruiting personnel! (NASA-JPL)

Explorer 1, the first US satellite plotting data from a network tracking station.

When JPL designed its early rockets the Friden mechanical calculators of the time could not do logarithms and Paulson used huge books of atmospheric densities compiled by the Works Progress Administration. Set up during the Great Depression and national austerity to get people working, it retrained in total almost nine million people so that they would be ready to re-mobilise the nation when times changed. Paulson's women took up those vast manuals and compilations and did they own calculations using this data taken to the next level.

Today, JPL employs almost 6,000 people with more under support services contracts or suppliers working off site. It has internships for High School students, post-doctoral and faculty students and offers job opportunities to many from those groups. JPL was also instrumental in setting up the Museum Alliance, a depository of materials and artefacts for educational groups and other space-related displays across the US. Its recognition extends to the film industry too, which frequently uses JPL and its staff for advice and background information on movie sets and depictions of space projects, real and imagined.

RIGHT The campus feel to the GSFC is enhanced by the spacious layout and leafy suburban district in which it is located, relatively close to Washington, DC, and the home of many science and environmental survey missions. *(NASA-GSFC)*

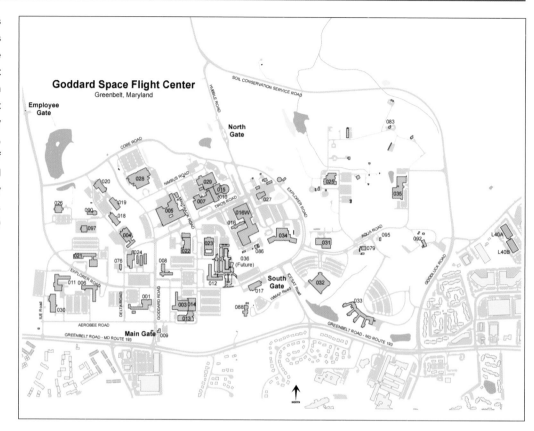

BELOW The Goddard Space Flight Center was the first new facility specifically constructed to support space science and associated research activities, although for a brief period it was considered the future home of human space flight programmes. *(NASA-GSFC)*

Not long after it was decided to set up NASA, a decision was made to build a dedicated space centre for research and for the development of satellites for scientific purposes. It had been on the cards for several months when Senator J. Glenn Beall of Maryland announced on 1 August 1958 that it would be located in his state on land surplus to the Department of Agriculture's Beltsville Agricultural Research Center. Despite opposition from Congress, believing such expenditure could wait, the then NACA went ahead and allocated internal funds, completing initial specifications by 16 September on what was then named the Beltsville Space Center, formally coming into existence on 15 January 1959.

With the national space programme coalescing around NASA, the Vanguard satellite team was to be transferred to Beltsville, although continued use of the Naval Research Laboratory's facilities for that project was essential while the new centre was under construction. On 1 May 1959 NASA announced this facility would be renamed the Goddard Space Flight Center (GSFC), after Dr Robert H. Goddard who, on 16 March 1926, became the first man to fly a liquid propellant rocket and it quickly began to assume an importance greater than that initially envisaged for it.

Goddard assumed responsibility for its first satellite project, Explorer 6 which was launched on 7 August 1959, and this numerical series of science satellites would continue until Explorer 35 which entered orbit on 19 July 1967 as the first of several Interplanetary Monitoring Platform series. From this inherited series came a wide range of projects, including the Orbiting Solar Observatory series from 7 March 1962 and the Orbiting Geophysical Observatory, the first of which was launched on 4 September 1964.

During 1959 staff levels for what was only a framework organisation before the facility was formally opened increased from 216 to 1,117, this figure inflated initially by the transfer of the Space Task Group (STG) responsible for the Mercury project from Langley to Goddard. STG became independent on 3 January 1961 which reduced the Goddard staff by 667 positions for a group which would form the core of the Manned Spacecraft Center.

While the dramatic launches were taking place from Cape Canaveral, exciting human space flight programmes were evolving through the Mercury, Gemini and Apollo programmes managed from Houston, giant rockets were being developed by von Braun's team at Huntsville and stunning discoveries were being made by missions run by the Jet Propulsion Laboratory, scientists and engineers at Goddard were quietly working to provide a broad and deep understanding of the space and near-Earth environment, of the Sun and of the radiation belts that framed the magnetosphere.

The first meteorological satellites were developed at Goddard, with Tiros 1 launched on 1 April 1960 providing the first global cloud-cover photographs. With further development the Tiros programme matured into the ESSA (Environmental Science Services Administration) series with the first launched on 3 February 1966. Contemporaneously, Goddard developed the Nimbus advanced atmospheric science satellites, first launched on 28 August 1964 and employed for studying a wide variety of

environmental issues covering polar ice cover, the Earth's radiation budget, the ozone layer and general weather.

Goddard was also responsible for the first Applications Technology Satellite, launched on 6 December 1966 as a communications research satellite but the series included some very advanced telecommunications satellites including the last, ATS-6, launched on 30 May 1974.

Placed in geosynchronous orbit, it was used for a communications experiment demonstrating the way satellites could be used to expedite education and disseminate lessons to remote rural communities in isolated locations, in this case the Indian subcontinent. In a UN-backed programme, ATS-6 was moved to beam directly down to India, schools programmes picked up on television sets provided to several thousand villages via antenna fabricated out of chicken wire by the students themselves. Programmes included instruction on how to use school chemistry sets to conduct soil tests to determine the best crops to grow in a particular area.

Communications research and development was a core technology application at Goddard, the first such being launched on 10 July 1962 when the Echo passive balloon was placed in space and inflated as a surface from which radio waves were "bounced" between transmitter and receiver. A much bigger satellite of the same kind, with an inflated diameter

BELOW **Dr Robert Goddard, the man after whom the facility was named which, unlike almost every other NASA site, retains the name it was originally given.** *(NASA-GSFC)*

LEFT The historic formal dedication, attended by Mrs Goddard in honour of her late husband, at a ceremony conducted on 16 March 1961. *(NASA-GSFC)*

of 135ft (41.1m), Echo 2 was launched on 25 January 1964. These very simple passive satellites were replaced with active repeater satellites able to receive and boost the signal before scrubbing it and re-transmitting it to a receiver. Relay was the first step toward that, launched on 13 December 1962, followed by Syncom on 26 July 1963.

From that early research came a deepening interest from the commercial telecommunications satellite industry with the launch of Telstar from AT&T on 10 July 1962, built by Bell Telephone Laboratories in a multi-national agreement with NASA, the United Kingdom and France's national telecommunications provider. The first international telecommunications service was offered following the launch of Early Bird on 6 April 1965 for the Intelsat network, a global venture with signatories from the national post and telecommunications agency of countries around the globe. Begun by the research work at Goddard and supported by other NASA field centres, this is one of the greatest contributions made by NASA in its more than 60-year history.

Goddard is also notable for having played a major role in expanding NASA's spheres of interest from ensuring American leadership in space science, technology and applications to the extended arm of government reaching out to engage with other countries in friendly and cooperative ventures. That began with the launch of the UK's Ariel I satellite on 26 April 1962 and on a joint US-Canadian satellite Alouette I on 29 September that year. Goddard also developed a research project with Italy and launched the San Marco I satellite on 15 December 1964, the first satellite built and instrumented in Europe and launched in the US by an Italian crew.

LEFT The Space Simulation Test Chamber number 3 which is used for vacuum and thermal tests on satellites and spacecraft designed for the harsh environment of space. *(NASA-GSFC)*

These international ventures were part of a "reimbursable launcher" programme which allowed foreign governments to fly civilian, scientific research and technology development satellites on American rockets for the net cost of the rocket and launch operations and without profit. This was opportune for countries which had not developed the capacity to launch their own satellites but it would bring frustration after one notorious occasion when the US government refused to provide NASA with a launch licence when the Franco-German satellite would have competed directly with a commercial US venture. It sparked such resentment that France became nationally committed to developing a European launch vehicle – Ariane.

Goddard has been responsible for more ground-breaking satellite programmes than any other NASA centre and while its mission is not as dramatic as some other facilities, in many ways it represents the very heart of what NASA is: basic research, doing things in space that the commercial world will not and pioneering in cutting-edge science and space applications. No greater expression of that is in the research work carried out with NASA and non-NASA satellites to expand the global database on the climatic changes taking place in the Earth biosphere.

Goddard has been responsible for environmental monitoring since the launch of Landsat 1 on 23 July 1972. Ever since, the Landsat series of satellites have played a major role in informing the global scientific

BELOW Developed at Goddard, the enormous volume of the Echo II communications satellite is evident by the scale of personnel alongside. *(NASA-GSFC)*

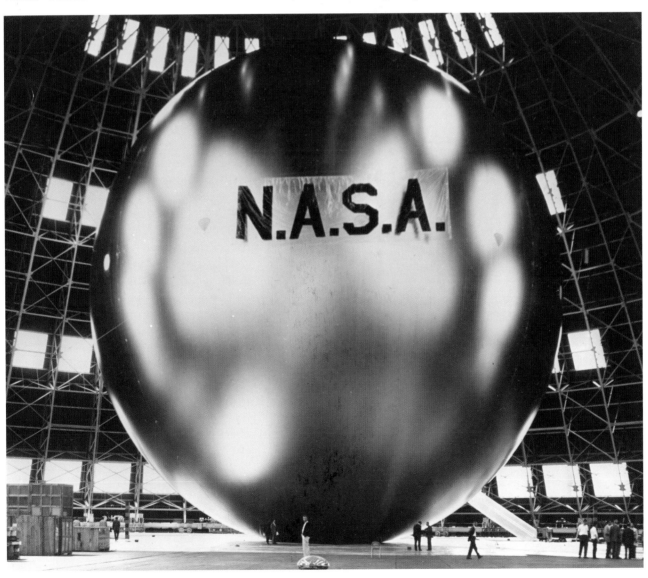

RIGHT The Goddard centrifuge for testing high-g loads experienced during launch and propulsive flight is an essential prerequisite for safe flight. *(NASA-GSFC)*

LEFT ATS-6 in Goddard's anechoic chamber to test antenna propagation patterns for communications equipment and transponders. *(NASA-GSFC)*

community about changing aspects of the Earth's environment and of the rapidly shifting balance of temperatures and the changing ecological balance. Known originally as the Earth Resources Technology Satellites (ERTS) programme, Landsat was a pioneer in providing free access to global environmental data through a series of ground stations in participating countries, many in the developing world and which benefit from timely data on water run-off, snow levels, the condition of surrounding seas and oceans and the advance and retreat of crop infestation.

This work provided an impetus to environmental studies much as the pioneering work in telecommunications had stimulated private industry to expand into the provision of commercial telecommunications satellite services. The development and operation of satellites and their services led to some rocky encounters between government agencies and Congress, which wanted to commercialise the work and for which a company was set up in 1985 under the name Earth Observation Satellite Company (EOSAT), a partnership between satellite builder Hughes and telecommunications company RCA. By this date Landsat had been transferred to the

RIGHT A false colour image from Goddard's Landsat 7 thematic mapper reveals geological data and information about the arboreal regions of the Island of Hawai'i. *(NASA-GSFC)*

National Oceanic and Atmospheric Administration (NOAA) but EOSAT ran into problems when Congress tried on several occasions to cut off government backing but support for the programme was formalised in 1992.

The most recent mission in this series, Landsat 8 was launched on 11 February 2013 but continued support for future satellites of this type is continually dependent on support from the White House and from Congress. NASA is no longer the prime agency rooting for Landsat since users now include many government departments, not least NOAA and the US Geological Survey, which sees in this programme a continuing value to the nation for its ability to secure sustained observation of national resources and the environment. But the work being pursued at Goddard is vital to maintaining a high scientific standard in the acquisition of global environmental data and that has never been more needed than now.

Today, the Goddard Space Flight Center employs about 10,000 people, most of whom are non-NASA contractor personnel. It still capitalises on almost 60 years of outstanding contribution to every essential element in space research. Its contributions to astrophysics, astronomy, instruments for planetary exploration – more of which have been contributed by Goddard than by any other centre – and to lunar science are legendary. It is the depository of a permanent archive at its National Space Science Data Center, first set up in 1966, and has a modest visitor information centre with a small rocket garden around the back, pride of place going to a full-scale model of a Delta launch vehicle, a rocket which has carried so many Goddard creations into space.

LEFT The Landsat 8 data-continuity mission observatory satellite undergoing tests at Goddard, this site being the home of NASA's environmental science survey work. *(NASA-GSFC)*

CENTRE The control room and its many consoles controlling the Hubble Space Telescope, returning daily hundreds of images supporting the work of scientists around the world. *(NASA-GSFC)*

RIGHT A view of the future, a telescope which will look back almost to the origin of the universe, as the James Webb Space Telescope comes together at a clean room at the Goddard Space Flight Center in anticipation of launch perhaps as early as 2021. *(NASA-GSFC)*

ABOVE NASA's Marshall Space Flight Center was inherited from the Army Ballistic Missile Agency, the dreams and designs of Wernher von Braun and his Operations Division having outgrown the requirements of the military. *(NASA-MSFC)*

This NASA facility came into existence on 1 July 1960, inherited from the US Army in a transfer of civilian personnel who had been instrumental in putting America's first satellite in orbit and who were already, when NASA formed in October 1958, developing a rocket at least as powerful as the Russian launcher which was already placing heavy payloads in orbit.

It was named after the legendary General George C. Marshall, US Army Chief of Staff under Presidents Eisenhower and Truman, Secretary of State and Secretary of Defense under Truman, architect of the Marshall Plan providing aid to war-torn Western Europe, winner of the Nobel Peace Prize in 1953 and President of the American Red Cross. The

history of the site chronicles the development of Army rocketry and the early interest taken in building rockets and missiles for military purposes.

Previously operating from Fort Bliss, Texas, US Army Ordnance officials sought a better location from which to carry out extensive development work. This had been initiated through tests of captured V-2 rockets with the assistance of the German rocket scientist Wernher von Braun. That work had been conducted at the White Sands Proving Ground, New Mexico.

On 15 April 1950 the Army transferred its rocket development work to Redstone Arsenal, Huntsville, Alabama, and took the von Braun team away from the spartan facilities at Fort Bliss to the more amenable surroundings of a more verdant southern state. On this date the Army set up the Ordnance Guided Missile Center (OGMC) from where sustained activity on producing a range of tactical, battlefield rockets began, largely with the help of the Jet Propulsion Laboratory (which see).

For six years the work focused on development of missiles such as the short-range Redstone (first launched successfully on 20 August 1953) and the Jupiter intermediate-range ballistic missile (IRBM), approved on 8 November 1955, until the establishment of the Army Ballistic Missile Agency (ABMA) on 1 February 1956. It was under the sustained support of von Braun that the Redstone could be used as a satellite launcher and that preliminary work was accomplished which enabled the von Braun team to successfully launch America's first artificial satellite on 31 January 1958. On 31 March 1958 the Army Ordnance Missile Command (AOMC) was formed. But von Braun had been working on a much more powerful rocket called Saturn upon which his team hoped to build a space programme.

By 1959 the future structure of the US nuclear deterrent was clear: an airborne strategic bomber force with gravity bombs; a land-based ICBM force eventually transitioning from non-storable propellant (Atlas and Titan I)

RIGHT Von Braun and his team, together with employees from US industry, created a rocket manufacturing facility supporting Saturn, Shuttle and the Space Launch System. His legacy is strong in the region with many families having settled there in the 1950s. *(NASA)*

to storable propellants (Titan 2) and rapidly to solid propellant (Minuteman); and a submarine-based deterrent based around the Polaris solid propellant missile. Using non-storable propellant, with a cryogenic liquid oxygen/liquid hydrogen upper stage, the massive Saturn I would have a capacity for placing 9 tonnes in low Earth orbit.

It had a thrust of 1.5 million lb (6,672kN) from eight rocket motors developed from existing missiles fed from a cluster of Redstone tanks around a central Jupiter tank structure. It was, in effect, a scavenged agglomeration of parts and motors from smaller rockets but it provided a lift-off thrust more than three times that of the most powerful US rocket launched to date – the Titan I. This meant it was far too big and powerful for any military purpose. Moreover, the Air Force could not find a reason for taking it over and with the lines of development already set, it was redundant to any military objective.

For NASA, however, it fitted a neat slot between the small- and medium-sized launchers it could use for satellites and space vehicles in the near-term and the Nova super-booster which was under consideration for the late 1960s or 1970s. At least that was the public story and one which has continued to populate NASA history books ever since. But there was a far different story which was itself fundamental to the establishment of the Marshall Space Flight Center.

Technical progress with the Saturn I design was going ahead well, the launcher anticipated as the propulsive stage for the Air Force's Dyna-Soar boost-glide vehicle. But Saturn was funded by ARPA and when Herbert York, the director of defence and engineering at the Pentagon, decided on 9 June 1959 to cancel the programme and shift Dyna-Soar to the Air Force Titan III it immediately sparked a reaction among supporters in the Pentagon and ARPA who feared an imminent retraction back from a heavy-lift capability. But Saturn I had no defined mission, either with the Department of Defense or NASA, and there was nothing tangible to justify its existence.

All parties recognised that America would need this heavy-lift capability eventually but the DoD was pressed financially and ARPA had much of its budget allocated to Saturn and

ABOVE The MSFC Rocket Park. Left to right: Hermes A-1; modified V-2; Juno II (foreground); Saturn I (background); Redstone; Juno I. (NASA)

was seeking a way out. York had a meeting with NASA boss Hugh Dryden on 16–18 September and closely examined the future for Saturn and the decision was reversed with the agreement that NASA would accept an offer for the agency to take it over. But NASA did not want to integrate such a vital part of its advanced missions planning and merely rely on a stated requirement to have the Army continue developing it. The only way out was for NASA to take it over.

BELOW Wernher von Braun and family in 1970 at the unveiling of a plaque acknowledging his contribution to the space programme and to the local community where large numbers of German families reside. (NASA)

1 July 1960 on which date it had 4,670 Army civil servants transferred to the space agency payroll, together with 1,840 acres (744.6ha) of buildings, from the Army Ballistic Missile Agency. Its first Director was Wernher von Braun, suddenly propelled from being head of the Development Operations Division, essentially a research and development group dependent on other ABMA offices for support, to boss of an entire facility responsible for the only vehicle which could give America equal lift capacity with the Soviet Union.

The day after President Eisenhower had signed the order the facility was renamed the Marshall Space Flight Center (MSFC) and the race to develop the Saturn I took on new momentum. Assembly of the first flight vehicle was completed at MSFC on 16 January 1961 and successfully launched on 27 October from LC-34 at Cape Canaveral (which see). When the Saturn I programme ended on 30 July 1965, ten flights had been completed without a single failure. Seeking a more powerful launcher, the Saturn IB had been announced on 11 July 1962, with the first flight on 26 February 1966. The last of nine flights took place on 15 July 1975, of which five had been manned by Apollo crews.

By transferring the Saturn project to NASA it would avoid inter-service rivalry and, with a first flight planned for 1961, give the space agency a much needed boost to short-term capacity. The move began with a high level discussion at the White House on 7 October 1959 which resulted in an agreement worked out on 20 October and approved by the President the next day. Along with the Saturn rocket, NASA would get von Braun's ABMA Development Operations Division.

On a presidential executive order signed by Eisenhower on 14 March 1960, this facility was activated under NASA management on

Development of the Saturn V, five times more powerful than the Saturn IB, was approved on 25 January 1962 as a result of the decision to adopt the Lunar Orbit Rendezvous mode for landing men on the Moon. The first of two unmanned launches occurred on 9 November 1967, after which ten were launched carrying manned Apollo vehicles, with the last, unmanned on 14 May 1973, lifting the Skylab space station into orbit. Throughout the period of development, many studies had been carried out by Marshall into further advanced versions capable of lifting greater payloads into orbit or to the Moon.

None of those came to fruition due to the

LEFT MSFC was an early convert to the use of water tanks to practise spacewalking operations. It built its own Neutral Buoyancy Simulator (NBS) which became pivotal to familiarising astronauts with procedures and developing new techniques applied in space. *(NASA-MSFC)*

retraction of the NASA budget and no evident programme for which a bigger rocket was required. In fact, the Saturn launch vehicles were deemed too expensive to sustain and the economic argument for reducing launch costs led to the flawed justification for building the Shuttle, to reduce those costs, and not for the justified argument that such a vehicle was required to build and assemble a modular space station, a role which it served with distinction. For the time being, however, the facility had its work cut out supporting Moon missions, and providing project management of the Boeing Lunar Roving Vehicle which would be carried on the last three Apollo lunar landing flights in 1971 and 1972.

Even before the Apollo programme had been completed, Wernher von Braun was moved up to headquarters on 27 January 1970 as Deputy Associate Administrator for Planning, his position as Director taken over by Eberhard Rees, who would remain there until 19 March 1973 when the last of the Germans retired. But the seeds for the future had already been sown as von Braun had urged upon NASA the need to build a permanent presence in space through an Earth-orbiting space station. The Marshall Center was already involved in managing the Skylab programme, using a redundant S-IVB stage fitted out as a habitable laboratory prior to launch. But the focus went deeper.

In 1969 Marshall managed Phase B definition studies into space stations envisaged as initially being occupied by 12 astronauts, growing to a base of 50 persons; Phase A studies were feasibility analyses with definition (Phase B) followed by Phase C, detailed design and manufacture and Phase D as completed assembly and flight operations. The space station studies would not get past the Phase B level; there was simply no money to build such an ambitious programme and Langley suffered the same fate with its MORL studies (which see). By this time NASA was committed

to a reusable Space Shuttle and while a station could not operate without a means to supply it, the Shuttle could be built first and then be used to assemble a modular station.

As the Shuttle programme evolved, and approval to proceed with development was given by President Nixon in January 1972, several NASA facilities became involved. While detailed design and management of the Orbiter would be the responsibility of the Manned Spacecraft Center, Marshall received management responsibility for all the propulsion elements: the three Space Shuttle Main Engines on each Orbiter, the two Solid Rocket Boosters and the External Tank to which the other elements would be attached.

New and innovative applications were introduced in the fabrication of components for the External Tank, including the introduction of lightweight materials as production matured, and friction stir welding which was also a technology shared between Marshall and the manufacturer, Martin Marietta. These technologies would go forward and be applied in later work on the ISS and also disseminated to industry as part of the NASA mandate to distribute detailed, non-proprietary engineering information to the general benefit of industry.

After Skylab, Marshall managed the Spacelab programme, a cooperative venture with the European Space Agency for a pressurised module, which would be fixed inside the Shuttle payload bay, and several open pallets for scientific instruments, which would be in the payload bay but exposed to the space environment. Management of the Spacelab programme gave both NASA and ESA experience in working cooperatively on international ventures. The first pressurised Spacelab module was flown on the ninth Shuttle flight launched on 28 November 1983, with the last of 32 missions on 11 February 2000.

Skylab and Spacelab gave Marshall the experience with manned modules carried inside the Shuttle but the decision to build what would later become known as the International Space Station gave the centre its first full role in the development of an independent orbiting research facility. By this time Marshall had been developing a wide range of major projects, including management of the Hubble Space

ABOVE The Destiny module now attached to the International Space Station was fabricated at MSFC. *(NASA-MSFC)*

RIGHT Space Shuttle Pathfinder (OV-098) being transferred into the Dynamic Test Facility, built for the Saturn I first stage. *(NASA-MSFC)*

Telescope (HST) with instruments provided by the Goddard Space Flight Center, launched on 24 April 1990. This was followed by the Compton Gamma Ray Observatory carried to orbit by the Shuttle on 5 April 1991 and Marshall also managed the Chandra X-ray Observatory, launched on 23 July 1999, bringing an end to the age of the great astronomical observatories.

Development of the International Space Station began in the mid-1980s with a series of proposals for a uniquely US facility supported by ESA, Japan and Canada but when the Cold War ended in 1991, the Russians accepted an invitation to abandon plans for a second Mir space station and join the international community, bringing their own extensive experience to the programme. Virtually every NASA field centre had a role to play in the ISS, and still does. But the experience with Skylab and Spacelab gave Marshall the job of managing development and fabrication of the Unity node attached to the adjacent Russian segment.

Unity, along with the Destiny laboratory module, the Quest airlock and experiment racks and facilities were built in the same facilities where the Saturn V first stage had been assembled. Boeing had fabricated the Saturn stage and was now working to fabricate the essential elements for the ISS. Marshall has a strong responsibility with the Payload Operations Center which is manned continuously, 365 days a years. In all, more than 1,600 investigations have been supported by this facility.

Today, Marshall has a central role in managing the Space Launch System, the heavy-lift launch vehicle capable of sending 95 tonnes into low Earth orbit and due to make its first flight in 2020. Technical challenges with this rocket have delayed the completion and there have been political interventions which skewed original plans for a vehicle of this class to support lunar landing missions which were cancelled and then reinstated. Much of the engineering and technology that went into the Shuttle programme has formed the base for the SLS, although the manner in which the programme is managed, as well as the engineering involved, is very different today.

Marshall has about 6,000 personnel on site, of which all but 2,300 are contractors utilising the extensive facilities at this site.

LEFT The general layout of facilities at the Marshall Space Flight Center, some of which have been dismantled in response to the changing requirements of the programme. *(NASA-MSFC)*

Responding to the major expansion programme authorised by the decision to take astronauts to the Moon, on 7 September 1961 NASA announced that it would be taking over a government manufacturing plant at Michoud, near New Orleans, Louisiana, for fabrication of the Saturn launch vehicles. The plant had been built in World War Two but had only come into widespread use during the Korean War. It was the first of two outstations of the Marshall Space Flight Center. Located on a water route to the Gulf of Mexico, it was perfect for transporting large rocket stages to Cape Canaveral.

The MAF was originally to support manufacture of the first two stages of the Saturn I and Saturn IB rockets and the huge S-IC, the first stage for Saturn V. The complex incorporated 890 acres (360ha) which accommodated 210ft^2 (19.51m^2) of engineering and office space, 1,959ft^2 (181.99m^2) of manufacturing space and 415,000ft^2 (38,553m^2) for storage, handling and maintenance. Initially, Chrysler settled in to

manufacture the Saturn I stages and Boeing used the facility for the S-IC. Mason-Rust did all the support work and kept the place furnished with services and plant.

The facility was originally designated as Michoud Operations on 18 December 1961 and assembly of the first Chrysler-built Saturn I first stage (S-I-8) began on 4 October 1962, with acceptance by NASA of the first two stages built by industry on 13 December 1963. Boeing activated the S-IC production line for the first stage of Saturn V on 22 October 1962 when it started tooling up and beginning component manufacture. Assembly of the first S-IC, the S-IC-D dynamic test stage, occurred in June 1965. On 1 July 1965 the facility was renamed Michoud Assembly Facility.

Michoud quickly filled up with personnel, rising to a peak of just over 12,000 civil service and contractor personnel by mid-1965, more than half being the Boeing personnel working on Saturn V's first stage. Slidell had a computer responsibility there with a total of just over 200 working the tele-computing support services. They provided the housing and facilities for the digital and analogue computers together with a computer centre. Slidell were located about 20 miles (32km) from Michoud and about 15 miles (24km) from the Mississippi Test Facility, now known as the Stennis Space Center (see next entry). Overall, staffing levels at the MAF would slowly decline as production of the Saturn vehicles ran down, dropping to around 10,000 by mid-1967.

But there was more work to be done as the Shuttle programme promised an extended production run for manufacture of the External Tank. Martin Marietta got the contract to fabricate those tanks, which went through a considerable evolution in weight reduction and design improvements, modifications which continually sought to reduce the weight and improve manufacturing processes. In all, Martin Marietta manufactured 136 ET assemblies, the first rolled out on 29 June 1979 with the last pushed out the plant on 20 September 2010. Lasting more than 30 years, it was

BELOW A general overview of the MAF where fabrication of the Saturn launch vehicle and External Tank for Shuttle took place and now the core stage for the SLS and elements of the Orion spacecraft are being produced. *(NASA-MAF)*

ABOVE Transitioning from Saturn I to Saturn V production (seen here) placed commercial manufacturers like Boeing in the centre of activity and helped develop a working relationship that exists at the heart of NASA operations today. *(Boeing)*

RIGHT The core stage for the Space Launch System impresses with its sheer size and gives scale to the fabrication facility itself. *(NASA-MAF)*

the longest single production run for any US government-funded space programme.

MAF got a false start on a new generation of work when it was gearing up to produce the Ares launch vehicle stages for NASA's Constellation programme. When President Obama cancelled that outright in 2010 that work stopped, only to resume in support of the Space Launch System when Congress insisted that the government would build a heavy-lift launch vehicle for future application to a variety of potential objectives. That work is now under way at the facility as the agency gears up for a new commitment to deep-space exploration with around 4,250 personnel on site.

RIGHT MAF is not all about big rockets and large stages. Here, a welder works on the Orion pressure vessel, an operation likely to give the facility a significant role in the future. *(NASA-MAF)*

ABOVE Where power comes alive! Stennis is the product of a name change but the role it had when it was known as the Mississippi Test Facility goes on. *(F. Morrison)*

The second satellite site reporting to the Marshall Space Flight Center was the Mississippi Test Facility (MTF), announced by NASA on 25 October 1961 as the site for testing the large rocket engines developed for the Saturn launch vehicles and for static testing of the assembled stages prior to delivery to Cape Canaveral for launch. Like the MAF, this site was also connected by water to the Michoud plant and the provision of this facility allowed access to warm-water ports and year-round activity free of extreme weather. Like the MAF it was served by the Slidell Corporation.

The decision to set up a national test site for large launch vehicle stages and their propulsion systems began when preliminary studies started in May 1961 and on 4 August that year a site evaluation committee was set up to examine 34 optional locations. Wherever it was, it had to be isolated from main urban conurbations, accessible by water and road, have good utilities already in place, local communities situated within 50 miles (80km) and an amenable climate for year-round operation – plus a lot of space around it serving as an acoustic buffer.

The selected site was located on the Pearl River in south-western Mississippi and close to the Michoud Assembly Facility. Technically, on 18 December 1961 its name was changed to Mississippi Test Operations (MTO) but that name never stuck and it was continually referred to as the Mississippi Test Facility, that name being officially purged in June 1963, a month after the first trees were felled prior to construction. Build-up was slow. MTO had only 24 personnel in June 1963, almost doubling 18 months later. On 1 July 1965 it got its name back as the Mississippi Test Facility.

Located in Hancock County, Mississippi, it

RIGHT A general view of the SSC test facilities. *(NASA-SSC)*

consists of a 13,428 acre (5,434ha) area for laboratory, engineering and office space and 368,000ft² (34,188m²) for storage and servicing. As initially built there was a complex consisting of two test stands for the S-II second stage of the Saturn V and a dual S-IC test stand together with associated utilities and support equipment. A navigable channel 8 miles (12.8km) in length provided access by barge from the Pearl River which delivered stages from the MAF where the S-IC was fabricated, and from Seal Beach, California, where the S-II second stage was assembled. Cryogenic propellants (liquid oxygen and liquid hydrogen) could also be delivered by barge.

Rocket motor development had largely been in the hands of industry, Rocketdyne having secured an early lead in large rocket motors with the powerful F-1 delivering a thrust of 1.5million lb (6,672kN) and the cryogenic J-2 a thrust of 200,000lb (889.6kN). Five F-1 would power the first stage of Saturn V with five J-2 in the S-II second stage and a single J-2 in the S-IVB third stage. Rocketdyne had their own test sites for engine development, test and qualification but MTF would be the only place where the assembled stages could be test fired prior to delivery to the Kennedy Space Center. While Marshall had been responsible for integrating these motors into proposed rocket stages, MTF would be responsible for testing the assembled configurations.

There were to be two sizes of stage test stand, forming the Rocket Propulsion Test Complex – the A-1/A-2 stands for the S-II Saturn V second stage and the dual B-1/B-2 stand originally for testing the S-IC first stage of Saturn V. The two A stands are similar, about 200ft (61m) tall and made from steel and concrete, built to withstand thrust levels of up to 1million lb (4,448kN) and temperatures up to 6,000°F (3,315°C). Each stand has its own dedicated infrastructure supplying liquid or gaseous hydrogen, liquid or gaseous oxygen, gaseous nitrogen and pressurising and purging gases.

The B-1/B-2 stands were built to withstand a maximum dynamic load of 11million lb force (4.99million kg) and consisted of dual assemblies, construction of which was completed on 13 February 1967. Like the

ABOVE Headquarters and general administration buildings at the facility carry almost 60 years of history as the place where the Saturn V came together in qualification of propulsion systems. *(NASA-SSC)*

A-1/A-2 stands, it has provision for propellants and fluids with modifications and changes to the service lines to adapt if for the cryogenic propellants employed to accommodate the SLS core for stage tests.

The historic arrival of the first rocket stage to reach the MTF occurred on 17 October

BELOW The A-1 test stand was built for the Saturn V S-II adjacent to the propitious supply of water! *(NASA-SSC)*

1965 after a trip lasting 17 days from the North American Aviation manufacturing facility at Seal Beach, California. Carried through the Panama Canal on the USNS *Point Barrow*, the S-II-T test stage was transferred to the barge *Little Lake* for a 45 mile (72.4km) journey up the Gulf Coast Intracoastal Waterway and the East Pearl River. The stage was successfully static fired for 15 seconds on 23 April 1966. The first Saturn V S-I-T test stage arrived on the barge *Poseidon* from Michoud Assembly Facility on 23 October 1966 and erected on stand B-2 where it was fired for 15 seconds on 3 March 1967.

BELOW Early days in construction of the A-2 stand as land clearance and foundations get under way. *(NASA-SSC)*

After the last Saturn V test firing, with S-II-15 on 30 October 1970, the A-1/A-2 stands were modified for tests with the Shuttle SSMEs and the first firing in support of that programme took place at stand A-2 on 19 May 1975. The last SSME firing was at A-2 on 29 July 2009, a test programme lasting more than 34 years. The stands are coming back into use for the J-2X engine which was first tested there on 18 December 2007. It is also used for testing the Aerojet Rocketdyne AJ26 motor, which powered the Antares rocket used by the then Orbital Sciences to launch their Cygnus commercial payload carrier to the International Space Station.

The last Saturn V test firing on stand B-1/B-2 occurred when S-IC-15 was run for 125 seconds on 30 September 1970, a month earlier than the last S-II stage firing on A-2. Neither stage would fly, both being retired to museum ownership, S-IC-15 now outside the Michoud Assembly Facility and S-II-15 at the Johnson Space Center. Of the other two Saturn V stages never launched, S-IC-14 went to the Johnson Space Center while S-II-14 is at the indoor Saturn V display at the Kennedy Space Center. Currently, the B-1/B-2 stands are being configured for static test firings of the core stage for the Space Launch System.

A new E complex was constructed in the 1990s for test firing small rocket motors and consists of four separate stands. In 2012 Blue Origin used E-1 for testing the thrust chamber of its new 100,000lb (445kN) thrust BE-3 liquid hydrogen/liquid oxygen motor but on 22 May 2014 an explosion during tests with an AJ26 caused major damage and further use of the E-1 stand went to Aerojet Rocketdyne for testing its AR-1 staged combustion cycle motor burning kerosene and liquid oxygen and delivering a thrust of 500,000lb (2,224kN).

The E-2 stand consists of two separate cells designed to support horizontal- and vertical-mounted engines respectively and has been modified to support engines operating on methane, some funds for this facility coming

LEFT The A-3 stand was constructed for the now-cancelled Ares launch vehicles when the Constellation programme was NASA's deep-space programme. *(NASA-SSC)*

from SpaceX and also from the Mississippi Development Authority. A wide range of components are capable of being tested here, including steam generators.

Additionally, the E-3 also has two separate cells, each for a specified thrust rating, and capable of supporting whole engine or component testing, such as injectors and various elements of hybrid rocket motors involving combinations for solid and liquid propellants.

The E-4 stand comprises four concrete-walled cells 32ft (9.75m) tall for testing 500,000lb (2,224kN) thrust rocket motors in a horizontal plane. It incorporates a signal conditioning building for facilitating data collection and a metal building with 12,825ft^2 (1,191m^2) of floor space incorporating a high-bay area, workshop and test control rooms. This facility is not completed and tenders have been issued for tenants to come in and utilise the vacant capabilities.

Another incomplete test stand is A-3, construction of which began after the decision to develop the Ares-I and Ares-V launchers for the Constellation programme. When the programme was cancelled the test stand became redundant. With a height of 300ft (91.4m), it was built to provide simulated altitude testing for large cryogenic rocket motors and was configured for a thrust level of 194,000lb (863kN) at 100,000ft (30km) altitude.

In addition, an H-1 test stand was prepared after the Department of Defense selected this site for fabrication of a test area supporting megawatt-class hydrogen fluoride lasers developed by the Ballistic Missile Defense Organization. The Space Based Laser was conceived in 2001 but little use was made of the facility. Beginning in 2007, British aero-engine manufacturer Rolls-Royce of Derby in the UK leased use of the site for tests which could not conform to acoustic restrictions at its home base.

For many space veterans it will always

be known as the Mississippi Test Facility but this site has seen two more name changes. On 14 June 1974 it became the National Space Technology Laboratories and changed to its present name from 20 May 1988 in memory of former senator John C. Stennis, a local politician and a strong advocate of the space programme.

BELOW The dual B-1/B-2 stands for the Saturn V first stage impress with size and provision for the powerful cluster of F-1 engines, now under modification for the SLS core stage. *(NASA-SSC)*

RIGHT The A-1/A-2 and B stands at Stennis are undergoing significant change for the new era of deep-space exploration and a return to the days of testing for a sustained production run of the SLS. *(NASA-SSC)*

Next to the Johnson Space Center, the Kennedy Space Center (KSC) is one of the most publicly reported places in America. Site of so many NASA launches, it was built to support the manned lunar landing programme announced by President Kennedy on 25 May 1961. Today, it sits on 219 miles^2 (567km^2) of Merritt Island north-west of Port Canaveral and not surprisingly is the most visited NASA facility of them all. Boasting a Visitor Complex and a rocket garden, it also contains the Shuttle orbiter *Atlantis* in an impressive display inside its own special building with full-size boosters and external tank inviting the visitor to enter.

The Kennedy Space Center is located on Merritt Island between the Indian River and the Banana River, with Cape Canaveral Air Force Station located on the spit running down between the Banana River and the Atlantic Ocean. The KSC Industrial Area is accessed across the Indian River via the NASA Parkway. Before the Kennedy Space Center came into existence, NASA conducted an extensive survey of potential launch sites for the massive Saturn V which would take men to the Moon. A joint NASA-DoD survey team was appointed in June 1961 and completed its work on 31 July when the conclusions were finalised.

Selection of Cape Canaveral was not inevitable. The Air Force had been there since 1 September 1948 when it took over the facilities of the Banana River Naval Air Station, an area on the east coast of Florida midway between Miami and Jacksonville covering 23 miles^2 (60km^2), which was chosen as the site for a missile test range. Prior to that, the area was a backwater of scattered fishing communities, untouched beaches and sparse houses – and a lighthouse which for several years into the Space Age remained a fixture reminding visitors of a long-gone past. Operating since July 1894, the current lighthouse is further inland than its predecessor completed in 1873.

Before selection of this site, from 1947 the Army had fired V-2 rockets from White Sands, New Mexico, where von Braun and his team were housed at nearby Fort Bliss. On 29 May 1947 a V-2 went awry and after launch it ended up in a cemetery south of Juarez, Mexico, one of several incidents which persuaded the Joint Chiefs of Staff to seek a more permanent site from where they could fire their rockets away from habitation. Cape Canaveral (at the time under a postal address as Artesia) was chosen for its remoteness, seclusion from prying eyes and proximity to the vast expanse of the Atlantic Ocean across which rockets with increased range could be fired in safety.

The site was officially designated as the Air Force Missile Test Center (AFMTC) on 30 June 1951 and would later mature into the launch point at the head of the Atlantic Missile Range. The first launch from Cape Canaveral took place from Launch Complex-3 (LC-3) on 24 July 1950 with the flight of Bumper 8, a modified V-2 missile carrying a WAC Corporal upper stage, followed five days later by Bumper 7, the last

BELOW ICBM row in 1959 at the dawn of the Space Age and the first launch pads emerging specifically for space launches. *(NASA)*

OPPOSITE As pads were added and facilities grew, when NASA moved in to occupy part of what was at the time Cape Canaveral Air Force Station, the site expanded north but the industrial area remained to the south. This map of Merritt Island facilities dates to about 1963 when it was planned that there would be three pads for LC-39. *(NASA)*

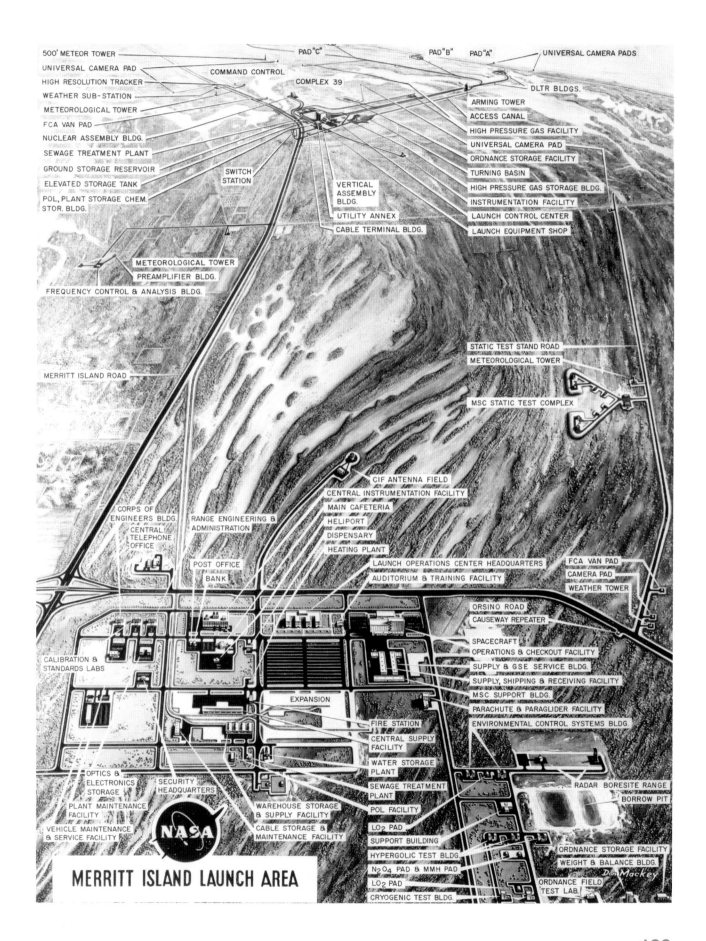

500' METEOR TOWER
UNIVERSAL CAMERA PAD
HIGH RESOLUTION TRACKER
WEATHER SUB-STATION
METEOROLOGICAL TOWER
FCA VAN PAD
NUCLEAR ASSEMBLY BLDG.
SEWAGE TREATMENT PLANT
GROUND STORAGE RESERVOIR
ELEVATED STORAGE TANK
POL, PLANT STORAGE CHEM.
STOR. BLDG.

COMMAND CONTROL
PAD "C"
PAD "B"   PAD "A"
UNIVERSAL CAMERA PADS
COMPLEX 39
DLTR BLDGS.
ARMING TOWER
ACCESS CANAL
HIGH PRESSURE GAS FACILITY
UNIVERSAL CAMERA PAD
ORDNANCE STORAGE FACILITY
TURNING BASIN
HIGH PRESSURE GAS STORAGE BLDG.
INSTRUMENTATION FACILITY
LAUNCH CONTROL CENTER
LAUNCH EQUIPMENT SHOP

SWITCH STATION
VERTICAL ASSEMBLY BLDG.
UTILITY ANNEX
CABLE TERMINAL BLDG.

METEOROLOGICAL TOWER
PREAMPLIFIER BLDG.
FREQUENCY CONTROL & ANALYSIS BLDG.

MERRITT ISLAND ROAD

STATIC TEST STAND ROAD
METEOROLOGICAL TOWER
MSC STATIC TEST COMPLEX

CIF ANTENNA FIELD
CENTRAL INSTRUMENTATION FACILITY
MAIN CAFETERIA
HELIPORT
DISPENSARY
HEATING PLANT

CORPS OF ENGINEERS BLDG.
CENTRAL TELEPHONE OFFICE
RANGE ENGINEERING & ADMINISTRATION
POST OFFICE
BANK

LAUNCH OPERATIONS CENTER HEADQUARTERS
AUDITORIUM & TRAINING FACILITY

FCA VAN PAD
CAMERA PAD
WEATHER TOWER

ORSINO ROAD
CAUSEWAY REPEATER

SPACECRAFT OPERATIONS & CHECKOUT FACILITY
SUPPLY & GSE SERVICE BLDG.
SUPPLY, SHIPPING & RECEIVING FACILITY
MSC SUPPORT BLDG.
PARACHUTE & PARAGLIDER FACILITY
ENVIRONMENTAL CONTROL SYSTEMS BLDG.

CALIBRATION & STANDARDS LABS

EXPANSION

FIRE STATION
CENTRAL SUPPLY FACILITY
WATER STORAGE PLANT
SEWAGE TREATMENT PLANT
POL FACILITY
LO2 PAD
SUPPORT BUILDING
HYPERGOLIC TEST BLDG.
N2O4 PAD & MMH PAD
LO2 PAD
CRYOGENIC TEST BLDG.

OPTICS & ELECTRONICS STORAGE
SECURITY HEADQUARTERS
PLANT MAINTENANCE FACILITY
VEHICLE MAINTENANCE & SERVICE FACILITY

WAREHOUSE STORAGE & SUPPLY FACILITY
CABLE STORAGE & MAINTENANCE FACILITY

RADAR BORESITE RANGE
BORROW PIT

ORDNANCE STORAGE FACILITY
WEIGHT & BALANCE BLDG.
ORDNANCE FIELD TEST LAB.

Don Mackey

NASA

MERRITT ISLAND LAUNCH AREA

ABOVE Hangar S became the home of NASA's first residential site for operational development of human space flight, a facility where spacecraft could be prepared for launch and where astronauts spent their last hours prior to flight. *(NASA-KSC)*

ABOVE Crew quarters in Hangar S where astronauts slept prior to launch. *(NASA-KSC)*

RIGHT John Glenn in the altitude chamber in Hangar S where the integrity of the pressure suit was tested before his flight. *(NASA-KSC)*

such legacy of German rockets from World War Two to be launched. This marked an end to the age of V-2 test firings and presaged a significant shift toward a broader range of military missiles and rockets which would find use much later when adapted for space research.

However, test firings with weapons continued from 25 October 1950 with the first launch from this site of the Lark surface-to-air missile (SAM), which had been developed toward the end of the Second World War as a defence against Japan's Kamikaze bombers. The missile was quite small, with a length of 6.25ft (1.91m) and a range of 34 miles (55km) but it was never much use. Test firings stopped on 8 July 1953 after 40 shots and Lark would be replaced by the Terrier SAM.

Activity really began to pick up pace with test flights of America's first pilotless cruise missile, the TM-61 Matador which had a range of 700 miles (1,125km). Classified as a tactical missile, Matador was 39.5ft (12m) in length with a 28.6ft (8.7m) wingspan. It was powered by a 4,600lb (20kN) thrust turbojet engine supplemented by a solid fuel rocket motor producing a thrust of 55,000lb (244.6kN) for two seconds to get the missile airborne. The first flight of the Matador took place at White Sands on 20

RIGHT The Mercury Control Center where the first seven NASA human space flights, six Mercury and one Gemini, were managed before moving to the new facility at Houston, Texas. (NASA-KSC)

January 1949 but tests shifted to Cape Canaveral from 20 June 1951 where 191 shots had taken place by 30 November 1956, followed by regular training launches until the type was retired in 1962. By this date 1,200 had been produced and deployed to several bases in Germany, South Korea and Taiwan. Test firings with an evolved Mace cruise missile started in 1956 and shifted to Cape Canaveral with the first firing from there on 29 October 1959.

On 29 August 1952 test flights began with the Snark cruise missile. Matador and Mace carried only fission warheads with a yield of 40KT but Snark was designed to deliver a 3.8MT warhead across a range of up to 6,330 miles (10,190km). It failed to live up to its promise and was cancelled in 1961, the last of the subsonic long-range cruise weapons replaced by the much more efficient ICBMs, support for which would occupy the Cape for many years.

Hard on the heels of tests with Snark, on 10 September 1952 the Air Force began test firings of the Bomarc surface-to-air missile, an

ABOVE Decked with the name Manned Spacecraft Center, Hangar S gets a visit from President Kennedy after the historic flight of John Glenn, the first American astronaut to orbit the Earth. (NASA-KSC)

LEFT Moved now to the Debus Center, the Mercury Mission Control room reimagines the setting from where the first NASA manned space flights were controlled. (Jud McCraine)

ABOVE The Space Mirror inscribed with the names of astronauts who lost their lives in the line of duty, mounted close by the KSC Visitor Complex. *(NASA)*

BELOW The home of Space Shuttle *Atlantis* where the Orbiter is presented in dramatic pose, one of the most popular displays at the Kennedy Space Center. *(KSC)*

integrated air defence system of which 1,140 were built and deployed between 1959 and 1972. The definitive missile had a length of 45ft (13.7m), a wingspan of 18.2ft (5.54m) and a range of 440 miles (710km). But as the total number of test firings of subsonic cruise missiles and rockets from Cape Canaveral began to build up, with 5 in 1959, 18 in 1951 and 41 in 1952, the first of the new generation of ballistic missiles began to emerge for trials which would increase in number over the next several years.

The biggest rocket launched from Cape Canaveral to this date, the first Redstone rocket was fired from Launch Complex 4A (LC-4A) on 20 August 1953. A direct evolution of the German V-2 rocket, it was arguably the most important product from the von Braun team at the Redstone Arsenal before the development

of the Saturn series of launch vehicles several years later. Designed to a requirement from the Army Ballistic Missile Agency for a short-range ballistic missile (SRBM), it would be capable of carrying a 3.5MT warhead to a maximum range of 201 miles (323km). When upgraded into the Jupiter-C test rocket and adapted to the Juno I configuration it launched America's first satellite into orbit on 31 January 1958. Modified further, the Mercury-Redstone would lift Alan Shepard on the first US manned ballistic space flight in the first Mercury shot launched on 5 May 1961, followed by Virgil Grissom on 21 July 1961.

Before the dawn of the Space Age, however, the last cruise missile programme of its era got under way at Cape Canaveral with the launch of a Navaho test round on 19 August 1955, the first of 12 in the X-10 series. The first full-up Navaho flight occurred on 6 November 1956. Navaho was a very ambitious programme, designed for vertical launch with two ramjet engines, each delivering a thrust of 15,000lb (67kN), lifted to its cruise trajectory by two rocket boosters each producing 200,000lb (890kN) of thrust. Designed to fly to its target at 1,990mph (3,200kph) it never achieved more than 1,550mph (2,500kph), theoretically capable of delivering a thermonuclear warhead to a maximum range of 4,040 miles (6,500km). When the first four all-up launches ended in failure the programme was cancelled on 13 July 1957, outmoded by the much more powerful ICBMs then coming along.

By 1957 America's first ICBM was ready to begin testing and new launch pads had been built to accommodate them, as identified on pages 128–188. The ICBM was considerably more powerful than the medium-, intermediate- and long-range rockets such as Redstone, Thor and Jupiter, calling for more sophisticated launch equipment and test procedures. They were seized upon as potential space launchers and in parallel with their evolution as an element of the nuclear deterrent, they also served to provide NASA and the Air Force with a capability to place satellites in orbit and send spacecraft to the Moon and the nearest planets. Their availability also sparked development of upper stages to increase their lift capacity and it was only through this combination that the early space programme came about.

Moonport USA

Little more than three years after NASA opened for business it was charged with mounting one of the biggest challenges in history: an open contest for putting the first man on the Moon, a gauntlet thrown down by President Kennedy to the Soviet Union at a time when most of the world believed that Russia was far ahead in the race for space. While the potential existed for selecting a different location for the launch of giant rockets to carry men to the Moon, Cape Canaveral was the logical choice because it was already hosting launch pads for flights with the Saturn I. Much bigger rockets would be needed to reach the Moon but all the industrial and administrative infrastructure was there already to support the early development stages of the Apollo programme before the bigger rockets began flying.

Management of the new facilities would drive the Launch Operations Center, then under the purview of the Marshall Space Flight Center, into a revised organisational box, creating an independent NASA installation. This avoided the burgeoning Manned Spacecraft Center from feeling that the launch facilities and operating protocols were in the pay of the Marshall Space Flight Center. A fellow scientist of Werhner von Braun from the V-2 days, Kurt Debus had been in charge of Saturn launch operations at Marshall and on 7 March 1962 it was announced that the launch facilities at the Cape would be independent, with Debus reporting directly to D. Brainerd Holmes at NASA HQ and not to Marshall.

Not for more than a year after the decision had been made did NASA decide the method it would use to get to the Moon, announcing in July 1962 on the Lunar Orbit Rendezvous (LOR) mode described on pages 35–36. That decision was based on use of the Saturn V rocket proposed by the von Braun team at

ABOVE The overall plan of Moonport USA, as it became unofficially known, was worked out by von Braun and Kurt Debus along with specialised construction teams from the US Army Corps of Engineers who managed the entire operation both at this end and down at LC-39. Here the Vehicle Assembly Building (VAB) is in the foreground with LC-39A in the distant right and LC-39B farther to the north. *(NASA-KSC)*

NASA-Marshall and this rocket required very special techniques for assembly, checkout and launch. And because of the sheer size of this rocket and the time it would take to prepare it for launch, it could not simply be assembled on the launch pad from which it would be fired.

Several methods were examined and the one chosen would allow the giant rocket, standing 363ft (110.6m) tall, to be assembled in a vast new Vertical Assembly Building (VAB), renamed

RIGHT The layout of the VAB accommodated a High Bay and Low Bay, seen here from the Crawlerway leading down to the pads, with the Launch Control Center to the left and set at an angle which placed it planar to the line between the VAB and the pads. *(NASA-KSC)*

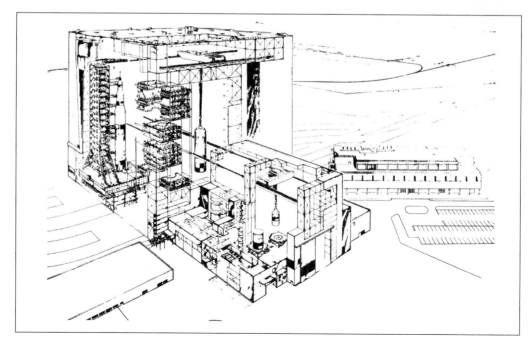

the Vehicle Assembly Building on 9 September 1965, capable of simultaneously processing up to four Saturn Vs. Inside the VAB each rocket would be assembled, stage by stage, on a Mobile Launcher (ML) which constituted the launch pad itself. The MLP would be placed on pedestals and when stacked together with the Apollo spacecraft on top, the launch vehicle would be carried on the ML to the launch site by a Crawler Transporter (CT), a flatbed with crawler tracks at each corner. At the launch site the CT would set the Mobile Launcher down on another set of pedestals, 22ft (6.7m) high and remove itself from the site.

Weighing 4,762 tonnes, the Mobile Launcher

(later known as the Mobile Launch Platform, or MLP) contained a Launch Umbilical Tower (LUT) with nine service arms supplying physical access, propellant, hydraulics, pneumatics and electrical power to the stack which would swing out of the way seconds before lift-off. Two arms were for the S-IC first stage, three for the S-II second stage, two for the S-IVB third stage, Instrument Unit and encapsulated Lunar Module and two for the Apollo spacecraft. The ML had a total height of 445.75ft (135.86m) to the top of the LUT, measuring 135ft x 160ft (41.1m x 48.7m) and 25ft (7.6m) high.

Although appearing to be a single monolithic platform, the ML was divided into two sections.

One, 135ft x 60ft (18.29m) physically supported the LUT while the other section, 135ft by 100ft (30.5m) provided a launch table for the Saturn V together with restraint arms and a 45ft^2 (4.18m^2) cut-out for the exhaust from the Saturn V, channelled down into the top of the flame deflector and along opposing sections of the flame trench. Inside, the ML consisted of two levels with 15 rooms on each throughout both levels. These rooms contained computers, checkout equipment, systems test equipment, propellant loading equipment, electrical equipment racks and engine hydraulic servicing units.

Some floors were mounted on shock isolators and the walls lined with acoustical fibreglass insulation to protect equipment. On top, four hold-down support arms constrained the rocket during the seven seconds after first-stage ignition with a pre-loaded toggle linkage released on a signal-to-launch commit, given only when the computers were satisfied that the five F-1 engines were up and running correctly and that nothing amiss was going on higher up. The hold-down arms were each 10.5ft (3.2m) tall and weighed 18 tonnes.

The LUT extended 398.5ft (121.5m) above the Mobile Launcher deck. At its base it measured 60ft (18.3m) by 111ft (33.8m), tapering to a 40ft (12.2m) square at the 80ft (24.4m) level. Two elevators operated at a speed of 600ft/min (183m/min) and had a carrying capacity of 2,500lb (1,134kg) servicing 17 work platforms and the swing arms. A hammerhead crane at the top of the LUT had a hook height of 376ft (114.6m), a traverse radius

ABOVE As tall as a 36-storey building, the Apollo 11 vehicle begins its move to the pad, displaying the segmented door through which the stack moves on to the Crawlerway. *(NASA-JFK)*

of 85ft (25.9m) and could move through a full circle. It had a maximum capacity of 22.68 tonnes, or 9 tonnes at maximum radius.

The nine swing arms extended from the LUT to critical levels on the rocket and the spacecraft and carried electrical and pneumatic feed from the ground to the Saturn V. Each weighed from 15.8 tonnes to 23.6 tonnes and varied in length from 45ft (13.7m) to 60ft (18.3m), all retracting within five seconds of the

BELOW A perspective view of the Apollo 11 rollout showing the rear of the Launch Control Center and the Mobile Service Structure (MSS) down the Crawlerway to the right of the picture. *(NASA-JFK)*

launch commit signal to release the hold-down arms. Construction of the first ML began in July 1963 at the Ingalls Iron Works in Birmingham, Alabama, and by March 1965 the structural frame for all three had been completed. Next came the installation of all the interior equipment and that was finished by May 1966. All three MLs were in service by late 1968.

Throughout the Apollo, Skylab and ASTP missions, ML-1 supported seven launches, ML-2 and ML-3 five each. Thereafter the name was changed to Mobile Launcher Platform (MLP). Extensive modifications for the Shuttle programme followed, the most obvious change being the removal of the LUT (the Shuttle was a little more than half the height of a Saturn V) together with its service masts and the hammerhead crane. Another change was the

BELOW The Mobile Launch Platform consisted of two areas, one for supporting the LUT (the lower part of this plan view) and the launch vehicle tie-down area with the central cutout to release the exhaust plume from the five F-1 engines on the Saturn V. Key: 1, dynamic support arms; 2, condenser; 3, stairwell; 4, umbilical tower column; 5, tail service mast; 6, vehicle engine chamber; 7, access hatches; 8, vehicle hold-down and support arms; 9, blast shield; 10, elevators; 11, deck-mounted cable enclosures; 12, environmental control system duct support. *(US Army Corps of Engineers)*

BELOW A view of the LUT on a Mobile Launcher at LC-39A showing the splayed lower section of the umbilical tower. *(NASA-KSC)*

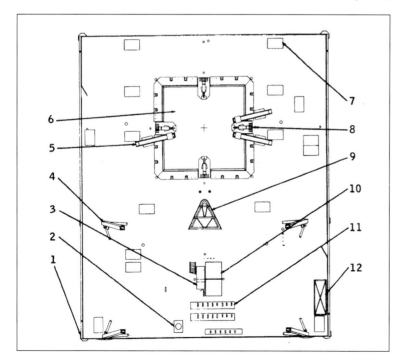

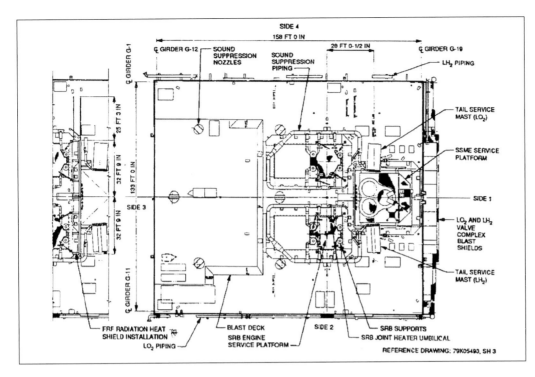

Diagram labels:
SIDE 4
158 FT 0 IN
CL GIRDER G-1
CL GIRDER G-12
SOUND SUPPRESSION NOZZLES
SOUND SUPPRESSION PIPING
28 FT 0-1/2 IN
CL GIRDER G-19
LH₂ PIPING
TAIL SERVICE MAST (LO₂)
25 FT 3 IN
32 FT 9 IN
193 FT 0 IN
SIDE 3
32 FT 9 IN
SSME SERVICE PLATFORM
SIDE 1
LO₂ AND LH₂ VALVE COMPLEX BLAST SHIELDS
TAIL SERVICE MAST (LH₂)
CL GIRDER G-11
FRF RADIATION HEAT SHIELD INSTALLATION
LO₂ PIPING
BLAST DECK
SRB ENGINE SERVICE PLATFORM
SIDE 2
SRB SUPPORTS
SRB JOINT HEATER UMBILICAL
REFERENCE DRAWING: 79K05493, SH 3

LEFT The Mobile Launcher was adapted for the Shuttle programme by complete removal of the LUT and rearrangement of cutouts for the exhaust from the three Shuttle main engines and the two Solid Rocket Boosters. *(NASA-JFK)*

replacement of the single central exhaust cavity with three exhaust holes, for the SSMEs and the two SRBs.

The two Crawler Transporters each weigh 2,721 tonnes and can carry a maximum load of 5,442 tonnes, for a combined moving mass of up to 8,163 tonnes, motive power coming from two 2,750hp (2,050.6kw) diesel engines driving four 1,000kW generators providing power to 16 traction motors. The motors turn four double-tracked crawlers spaced 90ft (27.4m) apart. Each of the eight treads is 7.5ft (2.28m) wide and 41.25ft (12.57m) long, with each tread weighting one tonne. Steering is controlled by two drivers in two cabs on opposing ends of the vehicle using an electronically controlled hydraulic system providing a minimum turn radius of 500ft (152.4m).

Because it has to climb an incline to reach the pad, it has a load levelling system consisting of hydraulic jacks which have a 6ft (1.82m) stroke for a maximum deviation of 2in (5cm) as measured at the top of the stack. The CT has a total length of 131.25ft (40m) and a maximum width of 114ft (34.75m). A separate power system provided AC power for the load levelling, jack, steering and ventilation as well as for the electronic systems and consisted of two diesel engines of 1,065hp (794.17kW) each and two generators of 750kW. The CT is

capable of 1mph (1.6kph) loaded and 2mph (3.2kph) unloaded.

Designed in 1962, NASA awarded the contract for the two Crawler Transporters to Marion Power Shovel Company in March 1963, sections being built in Ohio and shipped to KSC for assembly. The first was completed in 1965 and within two years both were in service. It lifted a Mobile Launcher for the first time on

BELOW The adapted Mobile Launcher for the Shuttle era with a cutback upper deck and reconfiguration of exhaust apertures and hold-down restraints. *(NASA-JFK)*

LEFT The Crawler Transporter (CT) provided mobility to the Mobile Launcher and to the Mobile Service Structure which was brought to the pad for servicing the Saturn V/Apollo launch vehicle. *(NASA-KSC)*

22 June 1965 and after some modifications it completed its first successful load-carrying run on 28 January 1966. Built for the Apollo programme, they were adapted for carrying the Shuttle on modified MLPs to LC-39. In 1985 a laser docking system was added, allowing the Crawlers to dock within 0.5–0.25in (1.27–0.63cm) of the "dead zero" mounting position at the launch pad.

Although assembly and stacking of the various stages and spacecraft elements took place in the VAB, there was a requirement to work on the vehicle at its launch site and for this a massive Mobile Service Structure (MSS) was constructed, brought to and from the launch pad by the Crawler Transporter. Weighing 4,444 tonnes, it stood 402ft (122.5m) above ground with a base of 135ft x 132ft (41.15m x 40.23m) and a top 113ft (10.5m) square. Design of the MSS consisted of a set of base trusses about

LEFT The overall dimensions of the giant CT, which at the time was the largest land moving object built. Refurbishment has prepared it for the Space Launch System which NASA is expected to begin flying from 2021. *(NASA-KSC)*

RIGHT The CT carrying the Mobile Launcher for the Shuttle era giving scale to the vehicles, the platform and the launch pad. *(NASA-JFK)*

22ft (6.7m) deep by 130ft (39.6m) square and a series of eight tower sections, each about 44ft (13.4m) high.

When completely assembled, the tower consisted of four upright trusses, side-by-side, each 335ft (102.1m) high, resting on a 22ft (6.7m) base truss 130ft (39.6m) square. Bracing members across the front and rear planes of the tower formed two additional lateral trusses and together the tower and base structure consisted of 1,458 separate pieces of steel weighing 3,488 tonnes connected at 416 joints. Five platforms were cantilevered from the forward face of the tower, opened as the MSS approached the pad and then closed around the Saturn V and spacecraft after the MSS was in place. Two elevators serviced the various work platforms, each carrying up to 16 people or 5,000lb (2,268kg) of equipment.

The MSS would be parked at a lay-by halfway between the VAB and the launch pad, well away from any debris should the rocket explode, only brought to the Saturn V/Apollo for technicians to work on the stack. It was first moved by a Crawler Transporter on 22 July 1966. The MSS was only used during the Apollo programme and was cut down after the last Saturn V launch, the Shuttle which followed having a completely different support requirement where the servicing tower was integral with the launch complex.

The entire layout of the facilities had been dictated by the enormous amount of energy contained in this giant rocket, which would have the explosive force of a small atomic bomb were it to detonate on the pad. For that reason the buildings and associated structures had to be dispersed so as to minimise recovery procedures in the event of a catastrophic disaster.

Central to the operation of the KSC complex was the VAB, at the time the largest building in the world. Originally, it had been planned to simultaneously prepare for launch six mighty Saturn V rockets in this building but as the design progressed it became obvious that launch rates beyond the capacity of the facility to simultaneously prepare four separate vehicles was not likely and, along with more than two launch pads, that layout was abandoned.

Construction of the VAB by the Army Corps

of Engineers began on 20 August 1963 after the area had been cleared the year before and more than 1,500,000yds^3 (1,146,900m^3) of soil had been deposited, raising it an average 7ft (2.13m) above sea level. The site was underlain with sand and compacted shell in the top 30–40ft (9.1–12.2m) over 80ft (24.38m) of compressible silt and clay. At a depth of 120ft (36.57m) there is a limestone shelf 3ft (91cm) thick overlying a stiff clay and silt bed on top of limestone bedrock at a depth of 160ft (48.77m).

The VAB was supported on 4,225 open-end steel pipe pilings, 0.375in (0.95cm) thick standing on the bedrock, a total length of pipes which would extend 128 miles (206km) if laid end to end. Each pipe has a compression load capacity of 90.7 tonnes to anchor the VAB and a tensile load resistance of 42.6 tonnes to prevent it being blown away in a hurricane, with 50,000yds^3 (38,230m^3) of concrete for capping the pile and providing ground floor slab. The primary skeleton of the VAB consists of 54,432 tonnes of structural steel in 45,000 separate sections weighing between 150lb (68kg) and 72,000lb (32.65 tonnes), one million high-strength bolts being used to fasten them together.

The truss system of the skeletal framework was laid out in multiples of squares 38ft (11.58m) on a side designed to survive blast pressures of up to 0.629lb/in^2 (4.3kPa) and suction pressure of 0.729lb/in^2 (5.02kPa). Erection of the steel framework began in

ABOVE Refurbishment of the CT with new propulsion units and electrical and electronic control systems, inside the VAB.
(Jacques van Oene)

ABOVE AS-503 at the
pad in preparation
for the Apollo 8
mission, with the MSS
approaching on a
Crawler Transporter.
(NASA-JFK)

leaves with a width of 76ft (23.16m) complete the vertical opening.

The doors can be operated separately and individually for environmental considerations, although there are 9,000 tonnes of air conditioning to keep the interior atmosphere under control. NASA's public relations team (not to mention the manufacturer of this equipment) were eager to point out that were it not for this installation clouds could gather inside and produce rain! This is not true, although the equipment does inhibit condensation, which would, in effect, drip off any vertical surface or ceiling. But it grabbed the headlines!

Essentially, the VAB consisted of two boxes: the High Bay, 526.75ft (160.55m) high by 518ft (157.88m) wide by 442ft (134.7m) long; and the Low Bay, 211.75ft (64.54m) tall by 442ft (134.7m) wide and 274.5ft (83.67m) in length. The two structures constituted a single internal volume of 129.482million ft^3 (3.664million m^3) and a total length of 716.5ft (218.37m) and the total height of the High Bay is almost as great as the Washington Monument.

Access to the stacked rockets inside this vast box was made possible by work platforms 60ft x 60ft (18.29m x 18.29m) and one, two or three storeys high. Manufactured outside the VAB, they were moved in when completed and raised to their appropriate levels but each was adjustable both vertically and horizontally at various elevations. Each would be cantilevered approximately 30ft (9.14m) when extended to wrap around the stack from opposing sides via semi-circular cutouts, sealed around the circumference of the vehicle with neoprene seals to prevent damage or spaces for tools to fall. Working on a platform cantilevered 40 storeys above ground was quite an experience and required a head for heights!

Personnel and light equipment trolleys could access various levels by 17 elevators, four of which were in the Low Bay. Elevators in the High Bay would provide access from the ground to the 34th level, 420ft (128m) above the floor of the VAB. Throughout the VAB, more than 70 cranes and lifting devices would provide leverage for moving things around, with the pre-eminent one being two bridge cranes with a 227 tonne lifting capacity serving the two High

January 1964 and the building was topped out in April 1965. The developed floor area totals 1,500,000ft^2 (139,350m^2) with 26 floors above ground level in the High Bay and three in the Low Bay consisting of lightweight reinforced concrete slabs 4in (10.16cm) thick.

The entire VAB is enclosed with 1,085,000ft^2 (100,796m^2) of insulated aluminium siding and 70,000ft^2 (6,503m^2) of light-emitting plastic panels. The siding stabilises thermal effects and helps to reduce acoustic pressure waves from the launch of Saturn V. The High Bay doors consisted of 11 separate leaves which provided a sufficiently large opening to allow the Crawler Transporter to roll out with the Mobile Launcher and Saturn V on its back. The space thus opened had a maximum height of 456ft (129m) comprising four leaves from the ground up which are 152ft (46.3m) wide up to the 114ft 34.75m) level. Above that level a further seven

Mobile Service Structure with the five platforms for accessing service levels on the launch vehicle and spacecraft.. *(Army Corps of Engineers)*

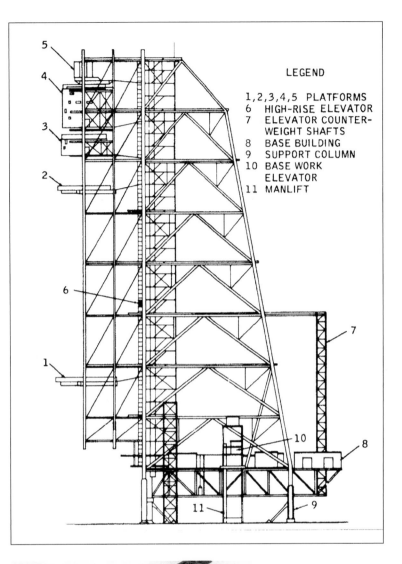

LEGEND

1,2,3,4,5 PLATFORMS
6 HIGH-RISE ELEVATOR
7 ELEVATOR COUNTER-
  WEIGHT SHAFTS
8 BASE BUILDING
9 SUPPORT COLUMN
10 BASE WORK
   ELEVATOR
11 MANLIFT

Bay assembly areas on opposite sides of the transfer aisle.

Each crane weighed 454 tonnes with a bridge span of 150ft (45.7m) and a hook height of 462ft (140.8m). A single bridge crane with a 154 tonne capacity and a hook height of 166ft (50.6m), ran along the transfer aisle between the High and Low Bay areas. Applicants for the job of driving these cranes were required to prove their expertise in moving massive weights with unprecedented skill by lowering the hook of a bridge crane to trap an egg on the floor without breaking the shell!

The concept of the Moonport facilities was unique to Cape Canaveral, the reception and assembly of various stages and spacecraft in the VAB unlike anything elsewhere, driven by a desire to prepare up to four Saturn Vs simultaneously and be in processing flow for up to two launches to occur within hours of each other. This defined requirements for the control of operations leading up to and including the launch into orbit. Instead of the traditional reinforced blockhouse, a Launch Control Center (LCC) was constructed alongside the south-east side of the VAB. This four-storey reinforced concrete building was 378ft (115.2m) long, 181.5ft (55.3m) wide and 77.2ft (23.5m) high, and was fabricated from pre-cast and pre-stressed material.

The LCC contained four firing rooms, each 80ft (24.6m) long and 140ft (42.7m) wide and two storeys high, containing 470 console positions for the small army of personnel required to count down and launch a Saturn V. The unique approach behind processing and launch of a Saturn V was the high level of automation, controlling the vast array of commands and sequencing required to prepare the various elements of the stack for flight. Nothing like this had been attempted before and it all came together at the LCC. The consoles informed personnel of the sequences programmed in to computers through several million pages of line-items, each a critical step

LEFT The AS-500F facilities checkout vehicle with the MSS in place, visibly displaying the two areas of the Mobile Launch Platform with the Saturn V at one end and the LUT at the other. *(NASA-KSC)*

ABOVE Firing Room 2 in the Launch Control Center (LCC), in this event during the pre-launch countdown demonstration test for Apollo 12. *(NASA-KSC)*

in a vast array of instructions and commands, worked out by engineers and technicians over the year preceding Saturn V flight operations.

A view of the two launch pads was possible through tinted windows consisting of 0.75in (1.9cm) thick laminated glass measuring 80ft (24.4m) by 22ft (6.7m) in each bay which allowed only 28% of the light to pass through and which deadened the sound from the first stage engines. The atmosphere inside the firing room was controlled by thermostats in the glass to prevent condensation and fogging by maintaining the temperature above that of the

external dew point, an area where 450 launch personnel were accommodated.

Construction started in March 1964 and was essentially completed by May 1965 although final finishing touches were not signed off before the end of that year, this building receiving the 1965 Architectural Award for Industrial Design of the Year. It was certainly aesthetic and contained within the working area of the VAB, with a total floor area of 213,900ft^2 (19,871m^2). The following year, the entire complex was selected by the American Society of Civil Engineers as the outstanding civil engineering achievement of the year. Designed by civilian architects and with the assistance of contractors, the entire site had been built by the US Army Corps of Engineers South Atlantic Division and is a lasting tribute to their skills and outstanding capabilities, still serving the nation more than 50 years later.

The Crawlerway was the specially constructed road which would support the Crawler Transporter and the Mobile Launcher as they carried the Saturn V to its launch pad. The route of the road was defined by a plan which initially anticipated the construction of five pads and the layout of the Crawlerway still bears evidence of the original idea to build three more pads farther north up the coast. After passing the turnoff where it leads to LC-39B, the Crawlerway turns east to carry the rocket directly to LC-39A where the road terminates.

RIGHT An external view of the LCC with added protection from hurricane-force winds. *(NASA-KSC)*

But if continuing on to LC-39B, the road turns again to proceed to its destination. But the Crawlerway was laid out to go north although that section was never built.

The distance between the pads was determined by the minimum spacing required to minimise damage in the event of a catastrophe befalling a Saturn V on either one, the two being 3.4 miles (5.47km) apart. The basic layout of the Crawlerway consists of two 40ft (12.19m) wide lanes separated by a 150ft (45.7m) wide median strip. It was laid down to withstand the 7,700 tonne mass of the Mobile Launcher and Saturn V rocket and averages 7ft (2.13m) thick. Work began in November 1963 under the same contract as that for the VAB and was completed in August 1965. Two-thirds of the way from the VAB, a branch extends to LC-39B and this section, 2.14 miles (3.44km) in length, was laid down under the contract for that pad. The distance from the VAB to LC-39B is 4.24 miles (6.82km).

The construction of the Crawlerway was determined by the loads and forces acting upon it and the four dual-tractor units of the Crawler Transporter was calculated to be 1,995 tonnes under normal operating conditions but when winds induced an imbalance this would increase to 2,449 tonnes and for this reason the Crawlerway was put down to withstand loads in excess of 12,000ft/lb (16,269.8n/m). A significant problem was that the Crawlerway would traverse different types of terrain which included swamp, dry land and sloughs with borings revealing that a satisfactory substrata existed 40–45ft (12.2–13.7m) below the surface and that at this depth it was reasonably compressible and consisted of interbedded clays, silts and silty or clay-type sands.

After unsuitable material was excavated, more than 3,000,000yds³ (2,293,800m³) of hydraulic sand fill was laid along the route, compacted with vibratory rollers and proof-rolled with a 90-tonne roller. The road itself consists of 3ft (0.9m) of graded crushed aggregate base course and 3.5ft (1.1m) of specially selected sub-base material. As the top surface on which the CT would operate, river gravel was placed to a depth of 8in (20cm) on curves and 4in (10cm) on the straight. The Crawlerway is 7.5ft (2.2m) above sea level

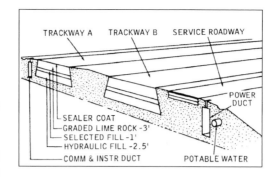

except for the 5% grade on the approaches to each pad. A 24ft (7.3m) wide service road runs parallel to the Crawlerway on the south side to LC-39A and the east side to LC-39B.

Utility and pipe lines are laid either side of the Crawlerway connecting the Launch Control Center and the VAB with the launch pads. Communication and instrumentation lines are placed in the ducts buried along the north side of the road, with as many as 40 ducts per bank, with repeater buildings at various intervals. High pressure gas lines are also situated on the north side of the Crawlerway supported on precast posts and piers, running from a high pressure gas facility alongside the Crawlerway. Power duct lines and potable water are along the south side of the roadway and where any lines or ducts pass under the road, access tunnels were built to withstand the prevailing loads.

Approximately halfway to the pads, a park position for the Mobile Service Structure was built on a 14ft (4.26m) thick mat approximately

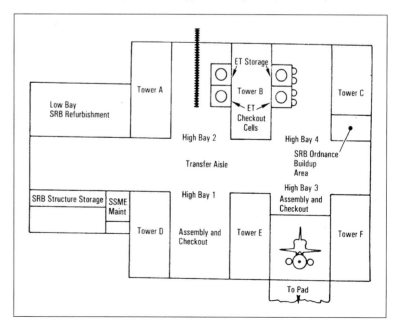

RIGHT The Orbiter Processing Facility (OPF) was constructed to receive returning Shuttle vehicles for safeing and also for the installation of large payloads mounted in the cargo bay with the vehicle in a horizontal position. *(NASA-KSC)*

BELOW A plan view of the OPF with two bays for working on Orbiters, one returning from flight and another being prepared for the next mission. *(NASA-JFK)*

200ft (61m) square containing 19,000yds^3 (14,527m^3) of concrete and 613 tonnes of reinforcing steel in the largest sizes available. The MSS weighed about 4,445 tonnes and the total load in the parked position when it was on the back of the CT was 7,167 tonnes. Four support legs held it in the parked position and with side struts fastened it was capable of withstanding winds of up to 125mph (201kph). With the end of the Apollo programme, the MSS was cut to pieces and some of the steel

used in the modifications to LC-39 for the Shuttle programme. Some of the steel was used to construct an observation tower from where visitors and tourists could view the two dominant launch pads and get a good view south toward Cape Canaveral.

Initially known as the Merritt Island industrial complex, ground was broken on 28 January 1963 with the first employees moving in during April 1965. This sprawling area lay to the south of the Moonport facilities and were the infrastructure that formed the backbone of operations. They remain as the core of the KSC industrial complex with all the facilities required to accept, prepare, process and deliver payloads, satellites and spacecraft to their respective launch pads. But it's difficult to avoid history at this most iconic of all the NASA facilities.

At 9.32am local time on 16 July 1974, astronauts Armstrong, Aldrin and Collins unveiled a plaque which commemorated the launch for the first landing on the Moon exactly five years prior to that moment. The plaque reads: "Men began the first journeys to the moon from this complex. The success of these explorations was made possible by the united efforts of Government, and Industry, and the support of the American people".

In retrospect, the Moonport facilities had been overbuilt, planned before it was known by which method NASA would reach the Moon and constructed when it was still believed that a programme of scientific exploration on the lunar surface would be a permanent feature of the US space programme. It was only the cost that had reduced from six to four the number of assembly bays in the VAB and it was the decision to build initially only what was essential for getting to the Moon at all that deferred the construction of up to five launch pads rather than the two that were eventually built.

After the last Saturn rocket had been launched in 1975, engineers descended on the launch complex once again, to modify and adapt it for the Space Shuttle programme. Beginning in 1976, two of the four High Bay areas in the VAB were converted for assembly of the Shuttle on a Mobile Launch Platform, the other two areas being adapted for processing and stacking the Solid Rocket Boosters and the External Tank. The north doors were widened by 40ft (12.19m)

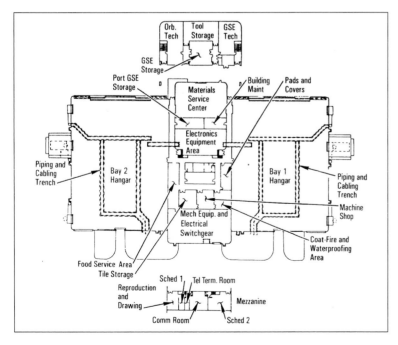

to allow entry for a towed Orbiter and its 78ft (23.77m) wingspan. Work platforms were added to Bays 1 and 3 where Shuttles would be stacked for launch and internal structural changes made to Bays 2 and 4. Changes to LC-39 pads are noted in the following pages.

While early Shuttle flights terminated with a landing at Edwards Air Force Base in California, where the landing strips were very much more accommodating for any anomalies in descent and landing, routine Shuttle operations would see the Orbiters return to KSC. The Shuttle Landing Facility (SLF) was a 15,000ft (4,570m) runway with a width of 300ft (91.4m) situated north-west of the VAB. Completed in late 1975, the runway slopes to the edge by 24in (61cm) from the centre and is grooved to prevent hydroplaning. The north-east corner supported a mate/demate device which would be used during the Shuttle era to lift the Orbiter off the back of a Boeing 747 after it had been returned from a landing at Edwards.

One significant change was necessary to accommodate the reusable nature of the Shuttle. It was expected that most landings would take place at the Shuttle landing runway, with the Orbiter wheeled around to a new Orbiter Processing Facility (OPF) alongside the VAB. Here the vehicle would be examined, serviced and made ready for its next mission before being wheeled around to the VAB for

vertical integration and rollout to one of the two pads at LC-39. It was also in the OPF that fixed payload structures would be installed while the Shuttle was horizontal. Initially, a third bay was built later north of the VAB, in effect the refurbished Orbiter Maintenance and Refurbishment Facility.

The three identical bays (OPF-1, -2 and -3) were each 197ft (60m) long, 150ft (45.7m) wide and 95ft (28.9m) high with an area of 29,000ft² (2,694m²) capable of processing two Orbiters simultaneously. The space between the two processing bays was connected by a low-bay area, 233ft (71m) long by 97ft (29.5m) wide and 24.6ft (7.5m) high. Beneath the common floor area were the electrical, electronic, hydraulic and pneumatic supply conduits together with gas supplies, all of which were connected to the VAB. A significant array of work platforms and access stands provided total access to the Orbiter along with three 27 tonne bridge cranes.

OPF-1 was closed following the rollout of *Atlantis* on 29 June 2012 and in 2014 an agreement was reached with Boeing for this facility to be used for supporting the Air Force's X-37B spaceplane. Later, a similar agreement was reached to embrace OPF-2 in the deal. Boeing also entered into a lease agreement for OPF-3 for supporting the company's CST-100 reusable manned space vehicle developed as part of a commercial partnership with NASA.

LEFT The Shuttle Landing Facility, a runway for returning Shuttle Orbiters and a landing strip for Orbiters being flown back from Edwards Air Force Base, now being used for the Air Force's Boeing X-37B winged reusable vehicles returning from space. *(NASA-JFK)*

Cape Canaveral launch pads

When the military arrived at Artesia's Cape Canaveral there was nothing with which to construct a launch site. Local materials scavenged from old beach houses, huts and a dressing room for swimmers sufficed to set up a makeshift pad for sending V-2 rockets into the stratosphere. For that was all America had to boast about: "liberated" German ballistic missiles left over after the Second World War and brought across from Germany. A single 100ft (30m) wide layer of concrete was poured on the sandy soil but there were no access roads to prevent cars, trucks and vans from getting bogged down to the axles in sand until a gravel route was laid and things slowly got better.

After the Air Force arrived in September

ABOVE A view of "ICBM row" as it was known at the end of the 1950s when it began to grow to accommodate test launches with a growing inventory of rockets and missiles, expanding on up the coast as the space programme grew as seen in this 1964 aerial shot. *(USAF)*

BELOW A map defining the limits of Cape Canaveral Air Force Station with the major launch complexes described in this section, also showing the Cape boundary south of the Kennedy Space Center. *(CCAFS)*

BELOW By 2016 when this map was prepared, many of the launch sites had underpinned the expansion of NASA and Air Force space activity had become redundant, as shown by active sites remaining, shown in red. *(USAF)*

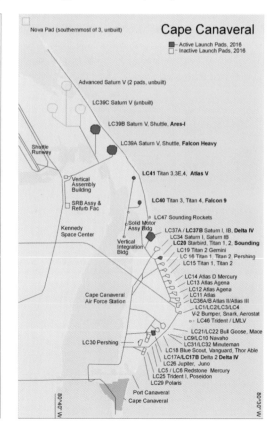

ABOVE Early launches from Cape Canaveral, for which the first few launch pads were purpose-built, were for military missiles and rockets such as the Bomarc, an anti-aircraft missile which formed the front line for the air defence of North America. *(USAF)*

RIGHT TOP Bomarc missiles erected from their horizontal position in protective enclosures. Canada procured the missile and cancelled its CF-105 Arrow, a highly advanced supersonic fighter which caused massive redundancies and brought some Canadian engineers down to the United States to work on the Mercury programme. *(USAF)*

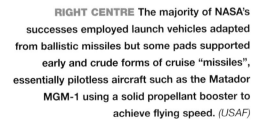

RIGHT CENTRE The majority of NASA's successes employed launch vehicles adapted from ballistic missiles but some pads supported early and crude forms of cruise "missiles", essentially pilotless aircraft such as the Matador MGM-1 using a solid propellant booster to achieve flying speed. *(USAF)*

RIGHT Potentially much more effective as a cruise missile, the Snark occupied a great deal of time and effort at Cape Canaveral, for little reward with a flawed system, again using booster rockets to get airborne. *(USAF)*

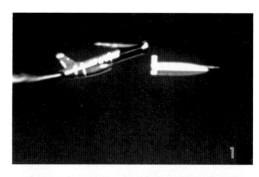

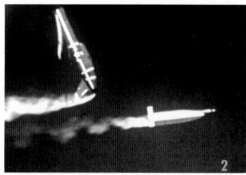

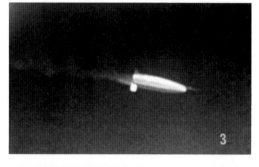

RIGHT Much of the testing with Snark was related to its guidance system, challenged by the projected range of more than 6,200 miles (10,000km). If it ever did reach that distance the forward nose section would detach and plunge to its target. The whole concept of the cruise missile was rendered redundant by intercontinental ballistic missiles such as Atlas and Titan. (USAF)

BELOW Impressive in size, redundant in concept, a Snark survives at the USAF Museum. (Greg Hume)

1948 it promptly set about establishing the Joint Long Range Proving Ground, formally approved as such by President Truman on 11 May 1949. The local Coast Guard opened up its small area at the Cape to missile use with construction work commencing on 9 May 1950, work undertaken by the Duval Engineering Company being completed on 20 June. These first two sites were designated Launch Complexes 3 and 4, the former becoming operational first. Scaffolding provided by painters sufficed to access the vertical V-2 rocket with plywood platforms laid across the shaky structure.

That swimmers' hut became the firing room, protected by sandbags piled across the front just 400ft (122m) from the rocket, and a line of trucks and trailers provided communications and links to tracking sites. The pad itself was a 100 x 100ft (30.5 x 30.5m) square, reinforced concrete slab 8in (20cm) thick. The firing room was mostly underground, a 20 x 20ft (6.1 x 6.1m) construction with periscopes for observation of the rocket.

Initially, Launch Complexes 1, 2, 3 and 4 were set up to support early flights and it was from these that the Bomarc, Matador, Mace, UGM-27 Polaris, Redstone and some X-17 launches took place, although they were not all activated at the same time.

In the following list of launch pads at Cape Canaveral, each is referred to as a Launch Complex (LC) followed by a number. Those familiar with military nomenclature will appreciate that a numerical sequence does not indicate a chronological order but rather a number slot assigned in a forward planning sequence. The date of first launch determines the sequence in which the following list is arranged, rather than acceptance, activation or operational readiness date, all of which are different, largely academic and of little historical relevance other than to indicate an intent.

However, compiling the list of pads in order of first launch allows the reader to see the growth before NASA was established and to appreciate the magnitude of launches which took place during the years of major expansion between 1950 and 1958 and those which were used to varying degrees by NASA programmes in subsequent years. The identified pad is followed by the date of first launch and its geographic coordinates in latitude and longitude.

A word on terminology. Name changes over time can be confusing. The Air Force now refers to a launch pad as a Space Launch Complex (SLC) but here we retain the original terminology of Launch Complex. Any references noted by readers, or seen depicted as such on visits to the Cape, to SLC with a number attached is the same as that identified here as an LC.

LC-3 (24 July 1950)
28.4655°N x 80.5349°W

This was the first of the initial quartet of pads built and was the first to be used when it launched Bumper 8 on 24 July 1950, the first launch from Cape Canaveral. A total of 77 launches took place from LC-3, the last on 12 August 1959, a Bomarc launch. When that programme ended LC-3 Bomarc facilities were used as the location of a medical facility during the Mercury programme. With toxic chemicals and corrosive agents in regular use, medical aid was a vital component of the individual launch pads.

LA (25 October 1950)
28.4650°N x 80.5370°W

One of the most prolific launch zones at the Cape, this Launch Area saw 340 launches of Snark Matador and Lark. Until the first Matador launch on 20 June 1951, flights had been exclusively with the Lark surface-to-air missile but tests with this cruise weapon continued intermittently with Lark launches until the first Snark from LA was sent off on 6 February 1953, an MX-775 precursor. The last Snark launch was on 10 December 1954 and a run of Matador launches set in before the last one of this type was launched on 1 June 1961.

LC-4 (10 September 1952)
28.4669°N x 80.5356°W

In terms of first use, LC-3 was to have been followed by LC-4, work starting on 5 September 1951 ready for the first launch of a Bomarc on 10 September 1952. LC-4 was similar to LC-3 but with one side of the concrete hardstand 200ft (61m) long with another area, 182 x 300ft (55.5 x 91.5m) surrounding the pad paved.

After this single launch of a Bomarc, the engineers moved in on LC-4 during 1953 to

ABOVE The first flight from Cape Canaveral occurred on 24 July 1960 from LC-3, a Bumper rocket supporting a research programme moved to Florida from the White Sands Missile Test Range in New Mexico and directed by General Electric. *(US Army)*

BELOW Launch Complex 3 was barely recognisable as the birthplace for Spaceport USA, little in the way of useable roads being available for moving the heavy rockets for which access was laid specifically by the Army Corps of Engineers. *(US Army)*

ABOVE Most of the initial design concepts for early Cape facilities came from the former German rocket engineers brought to the United States after the war, launch platforms like this at LC-3 being little more than a concrete base to which was attached a launch table. *(US Army)*

BELOW A makeshift gantry at LC-3 provides precarious access to the upper elements of the Bumper rocket which had a converted rocket, the WAC Corporal, as an upper stage.

adapt it for the Redstone. The first Redstone launch occurred on 20 August 1953 and this was followed by five more before a single Matador launch on 6 May 1955. On 2 February 1956 the first of 17 Bomarc launches occurred, the last on 15 April 1960 when the pad was retired.

As with other pads in this early series, LC-4 was used for aerostat ascents between 1983 and 1989. Historical interest in this pad is that it marks a signal change from the era of cruise missiles, defined then as airframes similar to swept-wing combat aircraft powered by jet engines and controlled by electronic autopilots. They were to be replaced by ballistic missiles of which the Redstone was the first to make its mark on pad configurations at the Cape.

LC-2 (18 February 1954)
28.4657°N x 80.5368°W

Completely out of numerical sequence, the next up to launch a rocket was LC-2 which, along with LC-1, was contracted on 18 October 1951, work taking almost 13 months but the Air Force only accepted the site in 1953. Specially constructed for the Snark cruise weapon, the first launch off LC-2 occurred on 18 February 1954, the last of 15 fired off the pad on 6 April 1960.

Used exclusively for Snark tests, it had no further use as a launch platform and went into disuse. After this, LC-2 was used as a helicopter pad during the Mercury manned flight programme and as a station for tethered balloon radar ascents in the 1980s.

LC-1 (13 January 1955)
28.4650°N x 80.5374°W

Long since deactivated, this was one of the four Snark pads originally set up for the winged cruise missile. LC-1 hosted its first launch on 13 January 1955 and supported 67 flights before the last was sent up on 5 December 1960. Like several locations around the Cape, from 1983 LC-1 was utilised for aerostat balloon missions and for several other activities, some of which are deemed classified. The last use of this facility was in 1989.

LC-6 (20 April 1955)
28.4407°N x 80.5726°W

Next up was LC-6 which was built to support

Redstone and Jupiter A flights. The complex was the first to use a mobile launcher concept developed by Dr Kurt Debus at Redstone Arsenal, engineers erecting a Missile Service Stand (MSS) to support Redstone launches. The MSS had a height of 135ft (41.1m), a width of 26ft (7.9m) and a length of 61ft (18.6m). Set on rail tracks so that it could be moved up to the missile on its pad, the MSS weighed 308,000lb (139,700kg) and supported a 13.6 tonne crane and four movable work platforms with an air conditioned work room, elevators and a standby power plant.

This amazing improvement over the facilities at LC-3 was well received by the workforce in this humid, mosquito-ridden swampland. Jupiter A was a modification of the Redstone to equip it for test firings of nose cones in the form of Jupiter-C and it demonstrated the use of Hydyne, a fuel which had more efficient combustion properties than the ethyl alcohol which had been utilised in the Redstone. LC-6 also supported Explorer and Pioneer flights as well as the Jupiter test shots. There were 43 launches from LC-6 between 20 April 1955 and 27 June 1961.

LC-5 (19 July 1956) 28.4394°N x 80.5733°W

This was the second in a paired pad shared with LC-6 but was first used more than a year after the first shot from that adjacent location when the Army launched one of its Jupiter A rockets to a height of 55 miles (90km) in the early hours of the morning. The range of 163 miles (264km) was a little long but this goes into history as the first Jupiter fabricated and assembled by Chrysler, all previous missiles being turned out from the Redstone Arsenal. The second launch, on 20 September, was a Jupiter-C re-entry vehicle test which performed as planned in almost all regards, except for an early shutdown due to an error in providing sufficient propellant in the tanks.

Several military flights followed before the launch of NASA's Explorer 3 on 26 March 1958, placing in an elliptical Earth orbit a satellite carrying science instruments largely associated with measuring the magnetic fields around the Earth. After an Army Redstone launch on 17 May, another Jupiter-C put Explorer 4 in orbit, followed by Explorer 5 on 24 August. Launched on 23 October 1958, the Beacon satellite

FAR LEFT Preparation at LC-5 on 16 May 1958 for the first launch of a Redstone rocket from Cape Canaveral the following day. *(NASA)*

LEFT Early successes were achieved with Juno II launches from LC-5, the space launch derivative of the Jupiter missile and another product from the von Braun group, this one carrying Explorer 7 successfully into space on 13 October 1959. *(NASA)*

LEFT Explorer 7 typified the research and data-gathering imperative driving the new agency, a satellite which obtained information about energetic particles in space and made the first measurements of the Earth's radiation budget which would begin the process of monitoring climate. *(NASA)*

was lost when the first stage failed. Beacon 1 consisted of an inflatable balloon structure but suffered structural failure due to the spinning payload just 2min 29sec after launch.

Some interesting flights followed, including Pioneer 3 sent up on a Juno II launcher on 6 December as an Army mission managed by NASA; this was one of several Moon probes which had been under preparation during the period NASA was formed. Although it failed to reach the vicinity of the Moon, achieving an altitude of only 70,000 miles (113,000km), it did discover a second radiation belt girdling the Earth. Following an Army Jupiter IRBM shot, Pioneer 4 was sent off on 3 March 1959 and sped past the Moon just 37,400 miles (60,200km) from its surface. This was the last of the four military Moon probes subsequently managed by NASA.

RIGHT LC-5 was used for the Mercury-Redstone ballistic flights beginning with MR-1 seen here on the pad prior to launch on 21 November 1960. *(NASA)*

FAR RIGHT The unmanned MR-1 launch from LC-5 got just 4in (10cm) off the pad before it settled back to its launch table, triggering the launch escape system designed to lift an astronaut free of a potentially disastrous situation. A spectacular failure, it was but one more dot on the learning curve to success. *(NASA)*

Further investigation of the radiation belts followed with Explorer 7 following a Juno II launch on 13 October before the first Mercury-Redstone shot (MR-1) took place on 21 November 1960. This was to have been the first unmanned test of the combination rocket and spacecraft scheduled to hurl an astronaut on a ballistic trajectory before orbital Mercury flights using the much more powerful Atlas rocket. The MR-1 shot failed when the motor ignited but immediately shut down, settling back on to the pad from a height of a few inches. A repeat on 19 December, MR-1A was a complete success.

On 31 January 1961, three years to the day after America's first satellite, MR-2 sent the chimpanzee Ham on the ballistic trajectory of the type astronauts would fly reaching a height of 155 miles (251km) before falling back to a splashdown in the Atlantic Ocean. In a bizarre twist of fate, von Braun wanted some modifications to the Redstone tested first in an unmanned shot before releasing the Mercury-Redstone combination to Alan Shepard, already selected to make the first ballistic space flight.

Mercury-BD was successfully flown on 24 March. Had it carried Alan Shepard he would have been the world's first spaceman, beating Yuri Gagarin by 18 days. In which case it is likely that the orbital flight of Gagarin on 12 April might not have had the knee-jerk reaction it did and in that event the Moon landing goal may never have been announced. As it was, Shepard flew MR-3 on 5 May and returned to a hero's welcome, clearing the way for Kennedy to make his famous announcement 20 days later.

The second manned Mercury shot on 21 July carried Gus Grissom on a repeat flight but when MR-4 returned he nearly drowned after quickly exiting a waterlogged cabin, flooded when the side hatch blew prematurely. This was the last of the 23 launches from LC-5, by which date LC-6 had already seen its last shot.

LC-9 (6 November 1956) 28.4520°N x 80.5562°W

Although we are not following a numerical sequence, suffice to note here that pads 7 and 8 were never built. The next pad to open for business was LC-9, employed for ten launches of the Navaho winged cruise missile between 6

November 1956 and 18 November 1958. The site had been accepted by the Air Force on 29 June 1956 but it was demolished in 1959 to make room for Minuteman complex 31 and 32.

Minuteman was the three-stage solid propellant ICBM which would provide the long-term solution to the Air Force requirement for an intercontinental missile which could be housed in silos. Successive versions improved range, performance and accuracy and the Minuteman ICBM remains in service today as America's only ground-based ICBM but a robust and highly

ABOVE The LC-5/6 blockhouse is now a museum which visitors can access via conducted tours. *(NASA)*

LEFT LC-9 supported test flights with the Navaho, the most advanced and complex of the winged cruise weapons developed alongside ballistic missiles, which for a variety of technical reasons had not persuaded all senior military officials that they were workable. *(USAF)*

LEFT A Navaho gets airborne from LC-9, propulsion provided by two Rocketdyne XLR83-NA-1 boosters each delivering a thrust of 200,000lb (890kN) with two central ramjet sustainer motors each with a rated thrust of 15,000lb (67kN). The programme was cancelled in 1957 but the rocket motor development paid dividends with later projects. *(USAF)*

accurate leg of the nuclear triad collectively involving land, sea and air deterrents.

LC-18A (8 December 1956) 28.4501°N x 80.5625°W

An interesting dual pad site specifically developed for launching the Viking sounding rocket and the Vanguard satellite launcher, effecting the transition from the ballistic era to the Space Age, distinctively separate from missile testing and weapons development which characterised many of the early sites. In fact the entire history of the Vanguard programme was played out from this pad as America's first official satellite programme, albeit unsuccessful in making the US the first country to launch a satellite and not even providing the first US satellite to reach orbit. But the chequered history of the Vanguard programme must not detract from the outstanding amount of information obtained by the satellites it did launch successfully.

The first launch from LC-18A was with a Viking rocket (#13) sent aloft at 01:05hrs in the first test of Vanguard components and related systems on a flight designated Vanguard TV-0 which carried a radio transmitter which was tracked to an altitude of 126.5 miles (203km). The second flight (TV-1) was another Viking rocket but with a prototype solid propellant Vanguard third stage as the second stage for this flight. Launched on 1 May 1957, it was a good flight, reaching an altitude of 121 miles (195km) and a range of 451 miles (726km). The third launch sent the first live Vanguard off on 23 October 1957 with inert second and third stages and this too was a total success. The first flight of a Vanguard with three live stages and a small 4lb (1.8kg) satellite failed during the launch attempt of TV-3 on 6 December when the first stage lost thrust and the vehicle collapsed down on to the pad.

RIGHT LC-18 supported the Vanguard satellite vehicle after previous flights with the Viking sounding rocket but the programme was plagued with technical difficulties. *(NASA)*

Coming just two months after Russia launched Sputnik 1 it was a deep blow to America's ambitions, directly triggering the order to von Braun for him to configure a Jupiter-C and to launch a satellite as quickly as possible, a move which resulted in Explorer 1 being launched from LC-5 on 31 January 1958 while Vanguard was struggling to catch up not only with the Russians but with the US Army! It did nothing for the project's reputation when another Vanguard attempt five days later failed after just 57 seconds of flight when TV-3BU suffered a malfunction to its control system and broke up.

But failure was not the preserve of Vanguard alone. On 5 March a second Jupiter-C launch carrying Explorer 2 failed to reach orbit while the first success for Vanguard came on 17 March 1958 when TV-4 was launched and placed the 4lb (1.8kg) Vanguard I satellite in orbit. Success at last preceded the failure of TV-5 on 28 April. This was the last of the "test" or development Vanguard launch vehicles. The next, sent up on 27 May, was designated SLV-1 indicating the first flight of the Satellite Launch Vehicle configuration, ostensibly now fully developed for a succession of science satellites. It failed when a disturbance upset the attitude gyro and the vehicle responded to incorrect pointing angles. The next two, SLV-2 on 26 June and SL-3 on 26 September, failed too.

Next off, LC-18A was a Thor-Able managed by NASA but for the Air Force Ballistic Missile Division. Launched on 11 October 1958, less than two weeks after the formation of NASA, Pioneer I was the first satellite launched by the new agency and was sent to an altitude of 70,700 miles (113,800km) before it fell back to Earth, re-entering the atmosphere two days later. But the small amount of scientific information it sent back was useful; not many space vehicles got that far!

Then, between 17 February and 18 September 1959 were a string of four Vanguard launches, all intended for orbit but two of which failed to make it; Vanguard was certainly not one of the more successful launch vehicles of the period but the last in the programme, after which LC-18A was turned over to the Air Force Blue Scout. This was a military version of NASA's first dedicated launch vehicles, the solid propellant

ABOVE A splendid view of the Vanguard on LC-18 prior to the attempt to launch the first American satellite on 6 December 1957. *(NASA)*

LEFT The explosion which resulted in the first satellite attempt brought criticism which only served to exacerbate frustration at America's lack of effective response to Sputnik 1 and 2, the last of which had placed a dog in orbit. *(NASA)*

multi-stage Scout small satellite launcher. Ten were launched between 21 September 1960 and 9 June 1965, after which the pad was retired and formally deactivated on 1 February 1967.

In 2016 the private development company Moon Express moved on to LC-18 (as well as LC-16) to carry out trials with its tiny lunar lander.

LC-17B (25 January 1957) 28.4458°N x 80.5656°W

The paired launch pads of LC-17A and B were among the early sites shaped by the dawn of the Space Age. Both A and B were constructed to support the PGM-17 Thor and were then adapted as that ballistic missile became the first stage for the Delta launch vehicle series. LC-17B was the first to launch a Thor which is why it is out of numerical order in this series.

The dual pad design benefitted immensely from the experience with earlier sites and had little relationship to the early pads which were largely to support cruise missile flights and were therefore set to send their vehicles off from ramps inclined so that booster rockets could get the winged missiles up to flying speed for turbojet or ramjet propulsion.

But these pads were different in that they were effectively highly efficient processing sites

ABOVE LC-17A and LC-17B displayed a more efficient and effective way of launching satellites and both served to send up a wide range of payloads. *(NASA)*

RIGHT From LC-17B, the first operational intelligence gathering satellite, GRAB, was launched on 22 June 1960 for the Naval Research Laboratory. *(NRL)*

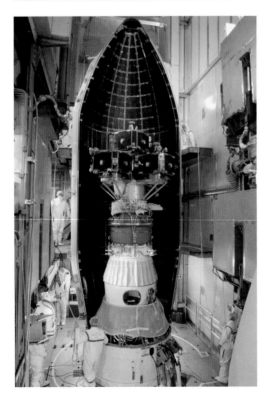

RIGHT The stack of five THEMIS satellites installed on top of the upper stage of the Delta launcher encapsulated by the payload shroud. *(NASA)*

FAR RIGHT Delta II on LC-17B about to launch the THEMIS II satellites. *(NASA)*

for quick turn-around and rapid post-launch damage mitigation. But Thor bequeathed a wide and expanding range of variants and types, each of which brought unique demands to the complex. Additional upper stages of varying length, longer stages, different booster motors – all required bespoke umbilicals, service towers, electrical, hydraulic and pneumatic feeder and liquid propellant fuelling provision for the core first stage and sometimes for the upper stage as well.

Over the more than 44 years that the LC-17B pad served the space-faring community, the site has been the departure point for some outstanding missions, including the Transit early navigation satellites, Courier military communications satellite, Telstar (the first commercial telecommunications satellite which inspired a UK pop hit from the Tornados and became the biggest US seller of its time), re-entry test vehicles in the ASSET series, Tiros weather satellites, OSO solar telescopes, ESSA environmental monitoring satellites, Intelsat international telecommunications satellites and many Explorer satellites.

Some truly historic missions included NEAR, a spacecraft launched in 1996 to the asteroid Eros which it orbited in 2000 before gently landing on its surface in February 2001. On 4 December 1996 a Delta 7925 launched the Pathfinder mission from LC-17B, the first spacecraft to carry a roving vehicle, the 25lb (11.5kg) Sojourner, to the surface of Mars. This was followed in 2003 by the launch of Mars Exploration Rover Opportunity, which began rolling across the surface on 31 January 2004 and was still trundling around as of May 2018.

Further iconic missions were to follow. On 3 August 2004, a Delta 7925H sent Messenger on its way to become the first spacecraft to enter orbit about the planet Mercury, in March 2011. On 12 January 2005, LC-17B saw the departure of the Deep Impact mission on a Delta 7925, aiming to study the characteristics of comet Tempel 1 through data from an impacting probe, which was accomplished on 4 July 2005. Another asteroid mission got off on 27 September 2007, entering orbit about Vesta in July 2011, departing for Ceres and entering orbit about that body in March 2015 where it remains.

On 7 March 2009, LC-17B hosted the

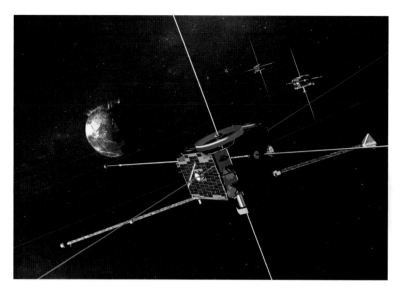

launch of the Kepler observatory on its mission to map the galactic star field seeking to identify stellar systems with planets. By the beginning of May 2018 it had identified 2,816 star systems encompassing 3,767 planets. The final launch from LC-17B was the dual payload of Grail A and B, spacecraft assigned to providing a detailed gravity map of the Moon from lunar orbit.

ABOVE The THEMIS satellite in an artist's depiction, one of five dedicated to studying space weather and the interaction of solar systems with the Earth's atmosphere. *(NASA)*

BELOW Another notable launch from LC-17B, the Telstar communications satellite sent up on 10 July 1962. *(NASA)*

The Air Force had transferred LC-17 to NASA in 1965 but reacquired ownership in 1988 for supporting the Delta II series. Both pads at LC-17 got an extended lease of life after the Challenger disaster of 28 January 1986 when it was decided to reopen production of conventional expendable rockets for payloads which were no longer allowed to the carried on the Shuttle, this action relieving the Shuttle programme of the urgency to launch "on time – every time", a factor in hastening to launch *Challenger* under less than optimum weather conditions.

In 1997 LC-17B was further modified to support launches of the Delta III rocket, first launched on 27 August 1998 in a failed attempt to place a commercial communication satellite in orbit; the guidance system went awry. The second launch of a Delta III on 5 May 1999 also ended in failure but the last Delta III, on 23 August 2000, was a success. Despite the fact that this vehicle had a unique configuration which required some changes to the LC-17B umbilical tower to accommodate the upper stage, the launch vehicle was dropped in favour of a new series, Delta IV.

The last use of LC-17B occurred on 10 September 2011, by which time 164 launches had taken place including the memorable missions mentioned above.

LC-22 (13 March 1957)
28.4610°N x 80.5398°W

One of two soft launch pads built to support Blue Goose missile launches. The SM-73

Goose was developed in the early 1950s as a delta-winged, ground-launched long-range decoy to simulate strategic bombers on a radar screen and confuse the enemy. The intention was to launch 50% of the total inventory within one hour of an alert, with the remainder one hour later. Goose would be launched up a ramp by a 50,000lb (222kN) thrust solid propellant booster motor and fly at Mach 0.85 on the power of a turbojet engine to a maximum range of 4,773 miles (7,681km).

Initial tests with Blue Goose began at Holloman Air Force Base in February 1957 but LC-21 and LC-22 were constructed beginning in 1956 and accepted by the Air Force on 26 February 1957. The sites were "soft" in that they did not support concrete hardstands, the firing van being located behind the inclined ramp. These missiles were advanced for their day and carried electronic equipment and transmission devices to fully simulate a B-47 or B-52 as they would appear to an enemy. The complex had two pads (22-1 and 22-2). The last of three Blue Goose launches from this pad occurred on 28 November 1958.

In fact the Blue Goose programme was never deployed operationally, other methods having been found to achieve the same result and the missile was cancelled on 12 December 1958. But after the last launch of this delta-winged decoy, significant modifications were made to the complex to support test flights with the Mace cruise missile, the first of which occurred on 29 October 1959 and the last of seven of this type on 24 June 1960.

The complex was vacated in 1964 and on 2 November 1970 it was reassigned to the Army for possible use as a test site for the Dragon anti-tank missile but this was dropped and the Air Force took it back on 23 July 1971 at which time it was dismantled and the blockhouse used as a store for supporting the Delta launcher programme.

LC-14 (11 June 1957)
28.4911°N x 80.5469°W

The pad was the first of four built specifically for the Atlas and as such represents the first ICBM launch site built anywhere outside the Soviet Union. Construction began in January 1956 when the US Army Corps of Engineers moved in with occupancy of the site completed a year later, acceptance coming in August after the initial launch. As constructed, the complex had a launch stand and a ramp with an associated blockhouse for control over the missile during pre-launch, launch and post-launch phases.

The launch stand was designed as the structural support for the missile in the horizontal position simulating the orientation it would have when installed as an ICBM inside a concrete coffin semi-buried in the ground to protect it from attack. This coffin concept was adopted for the test firings and the steel pad structure became the launch pad itself for LC-14 and the other Atlas pads at the Cape, approached up a ramp 92ft (29.5m) in length and 24ft (7.3m) wide. The pad hardstand was 57ft (17.4m) in length and 20ft (6.1m) wide and the rooms inside extended under the ramp itself, which was 22ft (6.7m) high where it joined the launch stand.

The stand was essentially a mounting platform for the hold-down and restraint system which would support the missile in the vertical position for launch, rotated 90° so that it could be mated to the aft end of the Atlas when the rocket was held in its horizontal transporter/erector frame on a trailer, which was reversed up the ramp by a vehicle which backed it up to the fixed restraint arms. When thus connected the assembly was rotated 90°, raising the rocket to a vertical position, after which the transporter/erector frame was lowered and

removed by the tractor. At ignition, flames and smoke would exit vertically from the base of the Atlas down into a void and turn 90° through a blast deflector to exit via an elliptical-shaped opening horizontal to the ground level.

The hardstand also supported an umbilical mast through which all the fluids, propellants, communications lines and electrical power supplies were fed from ground sources to the rocket itself. This stood 84.5ft (25.7m) above the stand. The blockhouse, or firing room, was about 750ft (229m) from the hardstand

ABOVE One of the more famous launch pads, LC-14 was used for the Mercury-Atlas launches, including sending the first four Americans to orbit the Earth into space between February 1962 and May 1963. *(NASA)*

LEFT Preparations underway at LC-14 for the launch of Gordon Cooper on MA-9, the last manned Mercury mission. *(NASA)*

and consisted of a concrete reinforced dome 60ft (18.3m) in diameter from where all the commands were sent and signals received from the rocket and its payload. To service the rocket prior to launch, a trapezoidal-shaped mobile service gantry, 154.7ft (47.16m) high, was moved on rails sideways from the parked position 300ft (91.4m) away from the hardstand and brought up to the rocket. Prior to launch it was moved back again to the parked position.

The propellants to fill the Atlas tanks were contained in large spherical vessels close by and delivered via over-ground pipes connected to the service mast. The liquid oxygen storage tank held 28,000 US gallons (105,980 litres) while the RP-1 refined kerosene tank held 16,000 US gallons (60,560 litres) of fuel.

The first five shots from LC-14 were with the Atlas A configuration, which only carried the booster engines and not the central sustainer motor, thus limiting the performance but considered prudent for such a radical design configuration. The first two failed, at 23 seconds and 32 seconds respectively, but the third launch on 17 December 1957 was considered a success, reaching a height of 74.5 miles (120km) and achieving a downrange distance of about 500 miles (804km). But this was only a small fraction of the true potential of the Atlas as an ICBM rocket.

To reduce weight, the prime contractor, General Dynamics, had designed the Atlas with paper-thin tank walls which also formed the external skin of the rocket, the structure pressurised with an inert gas to prevent it buckling under the weight and collapsing during storage and delivery. Under normal conditions the two booster engines would shut down about 2min 8sec after lift-off and separate by falling away from the aft end of the Atlas, secured together by a common attachment. The central sustainer engine would continue to fire until cut-off at 4min 48sec. Atlas was not a true multi-stage rocket but one which used a common set of propellant tanks for both boosters and sustainer motor and this was known as a 1½ stage concept.

LC-14 hosted two more Atlas A launches, both failures, before the first of three Atlas B, equipped with a central sustainer as well as booster motors was launched on 14 September 1958. The first operational Atlas D configuration got off LC-14 on 19 May 1959. Atlas D had been selected for manned Mercury missions to place astronauts in orbit and the first in a series of development flights took place on 9 September 1959. Planned as a test of the Mercury heat shield and dubbed Big Joe, it failed to achieve planned objectives when the booster failed to separate, which brought the spacecraft down short of its intended splashdown point.

On 26 November 1959 the first Atlas-Able configuration launched from LC-14 carried the Pioneer 3 lunar probe atop an Able solid propellant upper assembly. Able consisted of a liquid propellant stage with a solid propellant rocket on top, each to fire sequentially after launch by Atlas. Originally developed as upper stage elements for a Thor missile adapted for space flight, placing it on top of an Atlas ensured a heavier weight could be sent toward the Moon. It failed when a protective cover came away prematurely and doomed the attempt.

The next attempt, on 26 February 1960 saw the launch of an Atlas-Agena A, a more powerful upper stage which had been developed to carry a range of military payloads at first on the Thor first stage, not least the Corona spy satellites under the programme cover name Discoverer.

This launch carried the Midas 1 missile defence system satellite. While the Atlas stage performed well, difficulties separating the upper stage skewed the mission and the rocket fell back to Earth 2,500miles (4,022km) downrange. Midas 2, launched on 24 May, succeeded and became the first of its type in orbit.

After a research and development flight on 22 June 1960 LC-14 was turned over exclusively to the Mercury programme which began with the MA-1 flight on 29 July in which the capsule was to have been structurally tested during re-entry from altitude. The vehicle was destroyed at 3min 18sec when the adapter holding Mercury to the top of the Atlas failed. Another development flight followed on 22 October which was followed by three successful unmanned Mercury-Atlas qualification flights on 21 February, 25 April and 13 September 1961. The MA-5 flight launched on 29 November took the chimpanzee Enos on a 3hr 21min orbital flight, twice circumnavigating the globe.

Beginning with MA-6 on 20 February 1962 and ending with MA-9 on 15 May 1963, LC-14 hosted the four manned orbital flights concluding the one-man Mercury effort but it was not the end for support of NASA's human space flight programme. The decision to develop a two-man Gemini spacecraft to evaluate space walking, rendezvous and docking, and flights of up to two weeks required a target vehicle adapted from the successful Agena D, used for launching satellites into space and increasingly for sending spacecraft to the Moon and the planets. Under this plan, Atlas-Agena target vehicles would be sent up followed one orbit later by the manned Gemini flight for rendezvous and docking.

The first launch for the Gemini VI mission went off the pad at 10.00am local time on 25 October 1965 but it failed to reach orbit when the Agena refused to fire, sending itself into the Atlantic Ocean. This failure and the lack of an immediate replacement prompted the idea of launching Gemini VI after Gemini VII reached orbit on its planned 14-day flight and thus began one of the historic dual flights – the first time NASA had two manned spacecraft in orbit at the same time.

Next up on 16 March 1966 was the Agena target vehicle for Gemini VIII but after a

successful rendezvous and docking the crew of Neil Armstrong and Dave Scott (each to command Apollo Moon landing missions in 1969 and 1971 respectively) had to undock when their Gemini spacecraft got a stuck thruster and began to tumble, precipitating an emergency return to Earth. But the gremlins were still at work on 17 May 1966 when the Atlas set to launch Gemini IX's target suffered an engine malfunction, sending it down to a watery grave.

For just such an eventuality NASA had developed a slimmed-down version of an Agena target vehicle. Known as the ATDA (Augmented Target Docking Adapter) it had no propulsion but after launch on 1 June 1966 it was successfully placed in space by the Atlas alone and used for a rendezvous by the Gemini IX crew. But the shroud covering the ATDA had failed to completely separate, so preventing a docking. The last three Atlas-Agena target vehicles went off as planned: on 18 July for Gemini X, on 12 September for Gemini XI and on 11 November for Gemini XII.

LC-14 had directly supported NASA's first two manned space programmes, expanding space flight operations and providing a wide range of expertise. Preparation for the Mercury Atlas flights had been costly and extensive while modifications for the Atlas-Agena target flights necessitated gantry changes to accommodate the upper stage. After it was decommissioned, the complex began to fall into decay, the pace of the space programme had gathered momentum and somehow there never seemed

ABOVE The commemorative sign recognising the Mercury Orbital Flights located close to LC-14. *(NASA)*

to be enough money, time or people to preserve it as a national monument.

By the 1970s the red gantry and service tower were dismantled for safety reasons – the salty Cape air was not kind to metal structures – and by the 1980s it was sad to walk the ramp and look out over the shoreline from a two-storey height and see nothing, not even the remains of the gantry given to local rod-and-line fishermen to use as an offshore stand!

Long after Mercury, Gemini and Apollo had gone and the Shuttle had been flying for 17 years from a very different launch complex up the coast, the US Air Force 45th Space Wing moved in to begin some degree of minimal restoration funded by several aerospace companies and an army of volunteers from NASA, the Air Force and some civilian contractor personnel. A dedication ceremony was held in May 1998 where astronauts Gordon Cooper and Scott Carpenter were in attendance in addition to Betty Grissom, the widow of one of the three astronauts who died in the Apollo fire on LC-34 in January 1967.

Today, the site is a fitting tribute to the first two American manned space flight programmes and a large titanium structure sculpted to represent the "Mercury 7" – the seven astronauts of which four flew from this pad at the dawn of the Space Age. It marks a time capsule, not to be opened until 2464, 500 years after the end of the Mercury

programme, containing technical documents. Two kiosks flank the ramp to the hardstand. The adjacent launch complex blockhouse also contains documents and a photographic record of the facility.

LC-10 (12 August 1957) 28.4501°N x 80.5625°W

This complex supported just ten flights of the Navaho cruise missile and Jason and Draco sounding rockets after Navaho was cancelled, all launched between 12 August 1957 and 27 April 1959. Only the first flight was made with Navaho, the rest being a mix of Jason and Draco flights. Overall, the single Navaho launch was unsuccessful. After the successful boost phase the guidance system malfunctioned which prevented booster separation before reaching an altitude of 15 miles (24km). Pitch oscillations were overcome and the ramjets lit up to take the G-26 a distance of 174 miles (280km) at Mach 2.93. At that point the cruise weapon drifted off course and was ordered into a terminal dive by the range safety office, bringing it down in the Atlantic Ocean at a range of 264 miles (425km).

The next launch occurred on 15 August 1958 and was directly in support of a nuclear test using a Jason sounding rocket to send instruments to a maximum altitude of 430 miles (692km). Jason was a composite rocket utilising an Honest John missile as first stage, a Nike for second and third stages topped by a Recruit and a T-55 stage. It was developed specifically for measuring radiation levels in the high atmosphere. The last of six Jason flights from LC-10 took place on 2 September.

The last three flights from LC-10 started with a flight of a Draco on 16 February 1959, itself the first of only three Draco rockets ever launched. Built by Douglas, it consisted of two solid propellant rocket stages and was essentially a research derivative of the Sergeant missile, lifting instruments to an altitude of 11 miles (18km). The last flight from LC-10 – and the last ever Draco flight – was a failure.

LC-26A (28 August 1957) 28.4446°N x 80.5705°W

This complex was the first of two on a dual pad site with a common blockhouse and is

arguably second only to LC-39 in fame as it was the location from which the US Army launched America's first Earth-circling satellite. Construction of this dual complex began in 1956 and would continue to be used until 23 January 1963, LC-26A seeing fewer launches than LC-26B but with extended use over a longer time span. The site was deactivated in 1964 and on 20 November that year it was assigned as the US Air Force Space Museum, which has been open to the public since 1966.

The relatively small and contained museum includes the original blockhouse, several display and instrument consoles, the thick glass windows looking out across the two pads and an exhibit hall. Outside, there are displays of approximately 70 artefacts, including missiles, rockets and space launch vehicles.

The first flight from LC-26A was a Jupiter IRBM, notable for the first demonstration of separation from the thrust unit. It was followed on 31 January 1958 by the launch of Explorer 1, America's first satellite, on a Jupiter-C rocket hurriedly prepared for launch at the express order of President Eisenhower to counter the Vanguard failure the previous month. Paradoxically, it was followed on 5 March 1958 by the failure of another Jupiter-C carrying Explorer 2.

A succession of Redstone and Jupiter IRBM test shots followed, not all successful, some of which were NATO combat training launches

to qualify Italian and Turkish crews in a missile scheduled for deployment in Italy and Turkey. The 14th and last launch for LC-26A occurred on 23 January 1963.

LC-17A (30 August 1957) 28.4472°N x 80.5649°W

For details on the construction and history of this dual complex see LC-17B on page 136, which was the first pad at this site to go active.

ABOVE LEFT The launch of Explorer 1 on the night of 31 January 1958 from LC-26A, the first US satellite to reach orbit. (NASA)

ABOVE The LC-26A blockhouse that controlled the flight of the modified Jupiter-C test rocket and subsequent launches with vehicles of this type. (NASA)

LEFT Prior to launch, the forward end of the rocket is placed on top of the core stage, an integral part of the cluster of JPL solids which will carry the instrumented probe into orbit. (NASA)

RIGHT In place and ready for launch, the solids present a drum-shaped upper element to the Jupiter-C launcher. *(NASA)*

FAR RIGHT Sometime after the introduction of LC-17B, LC-17A came into use and with it the first launch by NASA, the Pioneer 1 satellite on 11 October 1958. *(NASA)*

Early launches from LC-17A were committed to supporting development of the Thor ballistic missile and then, from 17 August 1958 to the increasing numbers of space flights beginning with the launch of a Pioneer lunar probe, the first attempt by the United States to reach the Moon.

During the formation of NASA and before it had been inaugurated, the Army and the Air Force, both contested aspirants for the job of running the nation's space programmes, were each given permission to launch two Moon probes: the Army would launch Pioneer 1 and 2; the Air Force Pioneer 3 and 4. These were very small devices, the first weighing 83.8lb (38kg) and provided by the Air Force.

The Thor-Able rocket that launched this probe exploded just 77 seconds after launch and the probe was retro-designated Pioneer 0. Pioneer 1 followed from this pad on 11 October 1958 and reached a record altitude of 70,000 miles (113,000km), followed on 8 November by Pioneer 2 which failed to reach escape velocity.

As with LC-17B, a mix of re-entry test vehicles for the military and scientific satellites for NASA were mixed in the launch list with weather satellites, early types of navigation satellite and a range of interplanetary monitoring spacecraft – small satellites of the Sun placed to investigate the characteristics of the solar and interplanetary environment. But there were

international launches too, part of NASA's reimbursable launch programme in which the US government arranged to launch satellites made in and by other countries.

One such was the launch of the UK's Skynet 1A military communications satellite by Delta rocket on 22 November 1969. On 20 March 1970, another Delta launched NATO 1, a military communications satellite for the North Atlantic Treaty Organisation of which the US is the majority partner. Other Skynet and NATO launches followed. International launches included Palapa A1, a communications satellite for Indonesia, on 8 July 1976, the first of many for this country and followed by Palapa A2 on 10 March 1977, with Meteosat 1, a European weather satellite on 23 November 1977.

This period during the mid to late 1970s saw a rapid expansion in the commercial development of space projects and also in the broad application of satellites and space vehicles to human needs and requirements, improving communication in developing countries which allowed them to leapfrog the intermediate technologies still held on to in the developed world. Thus it was that higher capacity telecommunication systems in developing and third-world countries began to outstrip the general availability of such systems in the advanced nations such as the United States itself!

LC-17 was at the forefront of all this evolution but only because it was the bespoke complex for Thor, a rocket from which the extraordinary series of increasingly more powerful Delta launch vehicles emerged fulfilling the requirement for a burgeoning demand for commercial satellites. It was these that underpinned the explosive revolution in telecommunications, direct broadcasting, satellite TV and satnav systems, all built upon the expansive exploitation of access to space made possible by the Delta series of launchers from LC-17. LC-17A supported 161 launches, the last on 17 August 2009.

LC-26B (23 October 1957) 28.4433°N x 80.5712°W

A partner complex to LC-26A, it saw the launch of 22 rockets, the last on 24 May 1961 and was from the outset an important pad which saw the launch of a Jupiter IRBM (not to be confused with Jupiter A) on a research and development flight at its inauguration. Following that, a further 11 Jupiter IRBM launches took place before the pad was used for a Juno II launch on 15 August 1959 but this failed to place the Beacon satellite in orbit as intended.

A further four Jupiter IRBM launches took place before another Juno II launch on 23 March 1960, carrying the Explorer S-46 satellite, which also failed to deliver its payload to orbit. The next launch, on 3 November, successfully placed the Explorer 8 satellite in orbit, however. Three more Juno II satellite launches occurred, before the end of pad use on 24 May 1961, of which only one was successful.

LC-12 (10 January 1958) 28.4805°N x 80.5420°W

Along with launch complexes 11, 13 and 14, LC-12 was the second of four Atlas ICBM test sites built for the US Air Force to see a launch but was adapted later for several important space missions. The first 13 Atlas flights were assigned to testing of the Atlas D, the first operational variant of the ICBM, as well as some aeronomy flights for data collection in the high stratosphere and near-Earth space. These measurements were important for assessing the environment through which very long range flights would pass; not so much was known about the upper atmosphere at the time and there was a desperate need for more data.

On 25 September 1960, an Atlas Able was launched from LC-12 carrying a Pioneer P30 probe which was to have gone into lunar orbit, an attempt which had been postponed due to the backlog of flights from both the Air Force and NASA for access to the launch pads. When the pads had been laid down nobody had envisaged the demand from a burgeoning space programme. But the attempt failed when the Able stage ignited but rapidly lost thrust and decayed. Another attempt from the same pad on 15 November carrying the Pioneer P31 Moon probe was also a failure when the Atlas exploded 70 seconds after launch.

The next flight occurred on 23 January 1961, the final qualification test of the Atlas D ICBM, the 32nd of this series to fly and the 55th Atlas launch of all. A number of Atlas test flights had taken place from Vandenberg Air Force Base

BELOW One of four pads built to support Atlas launches from Cape Canaveral, LC-12 provides a visual record of the procedures used to back up the rocket to the hold-down restraint rig and elevate it to a vertical position for flight. *(USAF)*

in California but an increasing number of flights would be as space launch vehicles and activity at LC-12 switched exclusively to this work from 23 August 1961 with the launch of Ranger 1 on an Atlas-Agena B. This upper stage was an improved version of Agena A and was an early candidate for boosting satellites into orbit and the Moon before the Agena D entered service, an extended-length stage with multiple re-start capability.

Ranger 1 was the first in a series of unmanned probes designed to impact the Moon and to be equipped with a variety of different science packages including TV cameras to transmit pictures as it plunged toward the surface, allowing scientists to see objects only a few feet across. The Agena B stage pushing Ranger 1 to the Moon misfired and the spacecraft never got to its target. Neither did Ranger 2 launched on 18 November

1961 or Ranger 3 on 23 April 1962, both resulting from stage failures.

Launched on 23 April 1962, Ranger 4 reached the Moon and became the first American spacecraft to hit the surface three days later at 12:49:53 Universal Time. But a spherical hard-landing impact sphere designed to survive and provide seismic information and the TV cameras which should have sent back pictures could not work because the spacecraft began tumbling after separation from a perfect trajectory and the solar panels failed to deploy, starving its batteries of electrical energy.

Adding some sense of success, NASA's first interplanetary mission sent to fly close by the planet Venus was launched from LC-12 on 27 August 1962 just after dawn. But this only happened after an attempt to get its sister-probe Mariner 1 off the pad failed when the rocket had to be detonated by the range safety officer. Mariner 2 flew a perfect course to pass Venus on 14 December at a distance of 21,608 miles (34,773km) and return information about this cloud-shrouded world, indicating for the first time extraordinarily high temperatures from its carbon-dioxide atmosphere, findings which began the use of the phrase "greenhouse effect" when referring to a planetary atmosphere overheated by CO2.

Next off LC-12 was the Ranger 5 Moon mission on 18 October 1962, another failure to the Agena B which gave the spacecraft too much speed. Ranger 5 itself malfunctioned when the solar arrays failed to operate and the antenna lost lock on the Earth. After less than nine hours it was dead. The Ranger programme had been structured with a simple Block 1 design (Ranger 1 and 2), a more advanced Block 2 (Ranger 3, 4 and 5) equipped with survivable impact spheres and TV cameras, and a Block 3 which ironically were simpler but paradoxically more likely to succeed, carrying a battery of cameras and no capsule.

The failures of Ranger 1 to 5 brought scrutiny from Congressional space committees concerned at the waste in money and resources and a concerted effort was made to improve performance, largely through a more robust quality control system, replacement of some materials which were thought to have contributed to failure, and a simplified science

BELOW LC-12 supported many space launches, this D-series Atlas being used by NASA to launch the FIRE-1 re-entry test vehicle in a suborbital test on 23 April 1964. *(NASA)*

package. When the Ranger programme began, NASA was constructing a broad diversity of flights involving basic space science, the study of the Moon and the initial exploration of the nearest planets, Venus and Mars.

By 1963, as described in the general history of NASA earlier, all missions were focused on supporting the Apollo Moon landing goal and every mission was bent to that objective. The Ranger programme was supposed to do basic reconnaissance for the Surveyor soft-landers, two types of spacecraft together providing vital information engineers needed to make the final design of the lunar landing module and its supporting legs; scientists were still uncertain whether a spacecraft would disappear into an ocean of dust or merely penetrate a light covering across the surface. Ranger was supposed to decide what the surface was like by showing small rocks and boulders indicating only a light mantling of dust.

When flights from LC-12 resumed, Ranger 6 was launched on 30 January 1964 and the spacecraft flew a perfect trajectory to the Moon but failed to send any TV pictures home when the power system arced over. Rangers 7, 8 and 9 were launched on 28 July 1964, 17 February 1965 and 21 March 1965. They were all successful and returned thousands of TV images up to the point of impact, allaying the worst fears of scientists and giving engineers

the reassurance that they would not have to cope with significant amounts of dust.

Interspersed with these Ranger flights were some other historic launches. On 14 April 1962, NASA launched the FIRE 1 (Flight Investigation of Re-Entry) spacecraft for the Langley Research Center. Shaped like a miniature Apollo Command Module, the conical device was launched to high altitude by the Atlas booster and driven back down through the atmosphere by a solid propellant Antares rocket reaching a speed of 25,167mph (40,501kph). Radio data telemetered to the ground provided information about the environment on the heat shield and about the "black-out" period when communications were blocked by the plasma sheath built up as it entered the atmosphere. The second test, FIRE 2, was launched on 22 May 1965 and resulted in a velocity of 25,400mph (40,877kph), the speed which would be reached by an Apollo spacecraft returning from the Moon.

Another NASA programme, Orbiting Geophysical Observatory (OGO) saw its first launch from LC-12 on 5 September 1964, the Atlas-Agena B placing the OGO-1 satellite in a good orbit. It continued to operate until 1971. Launched from LC-12 on 7 June 1966, OGO-3 was a similar success. A more publicly memorable event occurred on 28 November 1964 with the launch of Mariner 4, the first spacecraft to fly past Mars, taking 21 pictures which transformed the way scientists viewed the planet. Mariner 4 flew within 6,118 miles (9,846km) of Mars on 15 July 1965.

Another "first" was chalked up by LC-12 when it saw the launch on 7 December 1966 of an Atlas-Agena D carrying the first of NASA's Applications Technology Satellite (ATS) series. ATS-1 was followed by ATS-2 on 6 April 1967 and ATS-3 on 5 November 1967, the last launch from LC-12. Before that, on 14 June 1967 another Atlas-Agena D sent the Mariner 5 spacecraft on its way to Venus, NASA's second spacecraft to visit this enigmatic world with a fly-past at 2,480 miles (3,990km) on 19 October, the closest yet of any planetary encounter.

Following the last launch, LC-12 was deactivated and for nine years remained dormant with only minimal maintenance. In December 1976 the pad area was completely dismantled and in September 2009 the blockhouse was demolished. There is very little there today to reflect its important place in rocket and space history.

LC-25A (18 April 1958)
28.4321°N x 80.5743°W

This was the first of four pads at LC-25, with only LC-25A and B initially surveyed in August 1956 for construction to support the solid propellant Polaris submarine-launched ballistic missile (SLBM) programme.

The overall contract for construction of the entire complex was issued on 19 March 1957 and LC-25A with access stand was completed by the end of the year. The single blockhouse was built to serve both A and B pads.

Between the date of the first launch and 5 March 1965, LC-25A supported a total of 60 launches of Polaris precursor development missiles and succeeding variants designated A1, A2 and A3 types, each with increased performance. The pad was dismantled in September 1969.

LC-18B (4 June 1958)
28.4490°N x 80.5618°W

The origins of this site can be found in the description of LC-18A (first use on 8 December 1956), and this adjacent facility was used for 17 test shots with the Thor missile, the last on 29 February 1960. After some modification and

rearrangement of support facilities and equipment it was employed for six US Air Force Blue Scout solid propellant shots between 7 January 1961 and 12 April 1962. The penultimate shot, on 1 November 1961, carried the Mercury MS-1 satellite which was to have qualified the global tracking and data collection network established for the one-man Mercury programme but the rocket failed 43 seconds after lift-off.

Along with LC-18A, the pad was deactivated on 1 February 1967 at a time when many changes were taking place at Cape Canaveral as the reality of budget cuts set in and a downsizing of NASA and Air Force plans coincided with a change in the type of missiles and rockets being tested. Later, the site was used for a de-mineralisation plant to service nuclear-powered submarines for reactor water.

LC-11 (19 July 1958)
28.4753°N x 80.5395°W

One of four launch complexes (11, 12, 13 and 14) built to support the Atlas ICBM programme accepted between August 1957 and April 1958. See LC-14 on page 139 (first launch 11 June 1957) for a description of the facilities design for all four. LC-11 was used for test flights with all production variants of this missile, including the initial flight tests with Atlas B, the first to carry a sustainer engine – the A series carried only the two booster engines for initial verification of the design concept.

The first five flights from LC-11 involved the Atlas B, the last on 4 February 1959 with the first of the production-level Atlas D on 29 July 1959. This was not the first D-series to fly, that distinction being claimed by LC-13 (which see). But on 18 December 1958, just over two months after NASA had been formed and with many launches failing and a general feeling that America was not catching up with the Russians, an Atlas B series was launched without a warhead or payload other than a radio transmitter. This was the first Atlas to go into orbit.

Known as Project SCORE (Signal Communication by Orbiting Relay Equipment), Atlas placed itself in an orbit of 922 miles (1,484km) by 114 miles (185km) and relayed to Earth a Christmas message from President Eisenhower previously recorded on tape. Purely as a propaganda exercise, the message had some

FAR LEFT A seminal success from LC-11 was the launch of an Atlas in Project Score on 18 December 1958, stripped down to minimum weight and with a carefully planned trajectory which placed itself in orbit and broadcast a message from President Eisenhower. *(NASA)*

LEFT LC-13 was also used for Atlas satellite launches among which was the launch of Mariner 3 on 5 November 1964, a mission to Mars which failed to reach its planned trajectory. *(NASA)*

BELOW Key to understanding the surface of the Moon and obtaining accurate mapping photographs was the Lunar Orbiter programme, five satellites scheduled to fly from LC-13 in 1966 and 1967. *(NASA)*

effect in raising morale regarding the Space Race, already attracting considerable interest among the press and the public. The Atlas rocket shell decayed back down through the atmosphere and was destroyed on 21 January 1959.

After ten Atlas D flights, tests with the E series began on 13 May 1961 and ended with the fourth such on 2 October 1961 with 12 F-series flights between 22 November 1961 and 28 October 1963. The last two launches involved an Atlas E on 25 February 1964 and another Atlas F on 1 April 1964. These test shots were not always designed for the rocket itself but to evaluate various re-entry bodies, particular types of trajectory and to more fully understand the re-entry environment in support of other existing rockets or future missiles.

The pad was deactivated on 13 June 1967 and the supporting service tower and associated structures were sold for salvage but the blockhouse for LC-11 was not demolished before April 2013.

LC-13 (2 August 1958)
28.4859°N x 80.5444°W

This was one of the four Atlas pads specifically built to support America's first ICBM and details on the construction of all four can be found at the entry for LC-14 on page 139 (first launch 11 June 1957). The first launch from LC-13

came only a few weeks after the operational inauguration of its sister-pad, LC-11 (which see). This pad would see flights of great importance in both the civil and military space programmes. After 13 Atlas B and D series flights, the last on 12 February 1960, an Atlas-Able conducted a static test firing from LC-13 on 15 February 1959 carrying a Pioneer P31 Moon probe but the rocket exploded.

Atlas D flights resumed on 11 March 1960 but the flight failed, as did three Atlas E launches in succession before the next successful launch on 24 February 1961, the first successful flight on an E-series in the Atlas programme. But two more E-series failed before the first of a number of successful launches on

26 May. An Atlas F was launched on 9 August 1961, the first from Cape Canaveral and then six more E-series, the last on 13 February 1962, the last of 18 E-series from the Cape.

Following this, some modifications were made to the pad to support orbital flights with the Atlas-Agena D, the first launched on 17 October 1963 carrying three small satellites delivered to different orbits by the restartable upper stage: two Vela radiation detection satellites and TRS-5, a small development satellite for space navigation systems, a very distant precursor of the GPS system of today. Two more Vela and TRS-6 went up on 17 July 1964, followed by Mariner 3, the first attempt to send a spacecraft on a fly-by mission to Mars, on 5 November. A failure in the shroud separation doomed that attempt but Mariner 4, launched from LC-12 (which see) was successful. On the next launch an Atlas-Agena D sent two more Vela and the ORS-3 radiation mapping satellite into orbit.

The most notable series of launches for LC-13 were the five NASA Lunar Orbiter spacecraft, which along with Ranger and Surveyor were the precursor reconnaissance missions to Apollo Moon landings. Five spacecraft of this type were launched between 10 August 1966 and 1 August 1967, all successful, providing the mapping detail required from which to plan the optional landing sites for the first manned missions. The Lunar Orbiter photographic archive is still regarded as one of the most important photographic archives of the Space Age.

Following those back-to-back launches, an Atlas-Agena D launched a Canyon communications intelligence satellite from LC-13 on 6 August 1968 followed by a second Canyon on 13 April 1969. These utilised the stretched Atlas with a 9.75ft (2.97m) tank extension for longer first stage burn and the satellite component remained attached to the Agena D in orbit where it deployed its 33ft (10m) diameter antenna. The orbits of the Canyon series were similar, those of Canyon 1 being 19,680 miles (31,680km) by 24,760 miles (39,860km) with a period of 23.9hrs – giving the satellite a slow migration around the planet with an orbital inclination of 9.9°, which means its ground path traces a path 9° north and south of the equator.

The next launch on 19 June 1970 was the first in the Rhyolite series, another highly classified project for the National Reconnaissance Office in the form of a signals intelligence gathering (SIGINT) asset. While Canyon operated in a near-geosynchronous orbit, the first Rhyolite (now known as Aquacade) operated from a highly elliptical path of 20,930 miles (33,685km) by 110 miles (178km). Others in the series operated from near-geosynchronous orbits. Alternating launches in a total of seven Canyon and four Rhyolite satellites saw out the use of SL-13, the last flight occurring on 7 April 1978.

LC-13 was deactivated on the day of last launch and although the gantry was declared a national historic landmark in April 1984 the mobile service tower was brought down in a controlled explosion on 6 August 2005 but the blockhouse was not demolished until June 2012. The site got a completely new lease of life in February 2015 when it was leased to the commercial rocket company SpaceX and renamed Landing Zone 1 (LZ-1) for the return to a vertical landing of the company's Falcon 9 rocket core, the first use of which occurred on 22 December 2015. LZ-2 was set up adjacent to take the two booster stages of the Falcon Heavy rocket launcher, and recovered on 6 February 2018.

LC-4A (3 September 1958) 28.4663°N x 80.5362°W

Constructed as a subset of LC-4, the complex supported three Bomarc test launches, the second on 21 October and the third on 27

January 1959. It was abandoned thereafter and has not been used for launches since, while providing storage and some support for aerostat flights.

LC-21 (25 September 1958) 28.4605°N x 80.5402°W

This complex incorporated two pads – LC-21/1 and LC-21/2 – built in 1956 and handed over to the Air Force on 26 February 1957 as test sites for the Goose missile, the last of three of which was launched on 5 December 1958. With the cancellation of the Goose programme the launch complex was modified for test flights of the Mace cruise missile, the first launched on 11 July 1960 and the last of 34 sent up on 17 July 1963. See LC-22 (page 138) for information on the fate of this and its adjacent pad.

LC-15 (6 February 1959) 28.4963°N x 80.5493°W

Along with LC-11, -12, -13 and -14 built to test and qualify the Atlas rocket, LC-15, -16, -19 and -20 were built for development of the Titan, America's second ICBM. Built to more conventional standards than the radical Atlas, which had pressurised tanks and 1½ stage configuration, Titan was slightly more powerful and was ordered into development largely as a back-up in case Atlas proved flawed. LC-15 was the first of those pads to see the first Titan launch.

The contract for construction of the four pads was signed on 23 April 1956 with construction commencing in February 1957 and completed on 1 July the following year. Like Atlas, the Titan pads used the horizontal coffins which they would be deployed in as operational ICBMs. Acceptance was signed off on 5 February 1959. The blockhouse was based on the Atlas design but with a second internal floor.

A series of 11 Titan I launches took place from LC-15, the last on 29 September 1960. Titan I used non-storable propellants, with liquid oxygen (LOX) having to be loaded immediately

ABOVE Completely transformed and repurposed, LC-13 is now used by SpaceX as Landing Zone 1 for returning core booster stages of spent Falcon 9 rockets.

FAR LEFT LC-21 was used for flights with the Goose missile and then, after modifications, with the Mace cruise weapon, an example of which is seen here. *(USAF)*

LEFT The first Titan I off LC-15, with dummy upper stage. *(USAF)*

prior to launch due to its cryogenic temperature. A more convincing deterrent was sought, one which would use storable propellants, dramatically minimising launch preparation time to just a few minutes.

The answer was Titan II, more advanced not only in that it had storable propellants but also in its guidance system and certain changes made to the operational flexibility for its nuclear mission. Support facilities for the Titan I had to be changed and modified for the Titan II, propellants now being nitrogen tetroxide instead of LOX and a blend of hydrazine known as Aerozene 50 instead of RP-1. These changes required modifications to the complex and the first of 16 Titan IIs was launched on 7 June 1962, the last on 9 April 1964. The complex was deactivated in March 1967, sold for salvage on 13 June 1967 and dismantled, the blockhouse being demolished between November 2012 and February 2013.

Titan II remained in service as an ICBM between 1963 and 1987, being retired on the basis of their age and the changing structure of America's nuclear deterrent. In their time, Titan II had the greatest throw-weight of any US ICBM, but there were never more the 60 in service at any one time. The development of the

Titan I and II ICBMs underpinned a significant expansion of their use into satellite launchers, later versions with strap-on boosters to provide a truly heavy-lift capacity for large payloads consisting of classified national security satellites and probes for NASA.

LC-19 (14 August 1959) 28.5069°N x 80.5542°W

Contracted at the same time as LC-15 (which see), construction began on 18 February 1957 and was completed in July 1958 with acceptance by the US Air Force on 16 June 1959. LC-19 supported 15 Titan I launches, the last on 29 January 1962. The first launch attempt was fouled by a fault with the hold-down arms which prematurely released the vehicle for flight before total thrust build-up, causing the Titan I to fall back on the pad. Not until 2 February 1960 did the next attempt take place, which this time resulted in a successful flight.

Most launches supported development of the Mk 4 re-entry vehicle (RV) in a test programme designed to qualify it for application with Atlas E/F and Titan I ICBMs. Produced by the Avco Corporation (the people who made the Apollo heat shield), the Mk 4 had an aluminium inner frame supporting an ablative external coating to protect it during re-entry. The Mk 4 contained a W-38 nuclear warhead with a yield of 3.75MT, approximately 250 times the explosive yield of the Hiroshima bomb.

A notable launch occurred on 25 July 1961 when a Titan I was launched during the early afternoon utilising the new all-inertial guidance system and became the first to complete a full range test, demonstrating a strike distance of 5,000 miles (8,040km). Titan in general had a good track record for its time. When the last Titan I fired off from LC-19 it was the 47th and last Titan I development flight to fly from Cape Canaveral, of which 72% had been fully successful and only 8.5% had been total failures.

With the transition from the Titan I to the Titan II (see LC-15, page 151) there was a requirement to adapt the basic Titan launch pads to the new rocket and while that was taking place, NASA assumed responsibility for adapting the modifications for LC-19 to the very specific needs of the Gemini programme. Begun in early 1962, the Gemini programme

RIGHT A fully configured two-stage Titan I. *(USAF)*

would conduct research into rendezvous and docking, long duration space flight and spacewalking – operations which would have to be developed for Apollo Moon missions and which could not be provided by the limited capabilities of the one-man Mercury programme.

The launch vehicle for the two-man Gemini was to be a man-rated Titan II, known as the Gemini Launch Vehicle (GLV), two unmanned qualifying flights and ten manned missions, all of which would fly from LC-19 and complete all flight activity at this launch site. The GLV was a major improvement over the standard Titan II and had several unique technology capabilities including a suppression system for preventing vertical oscillations during ascent and a malfunction detection system to help the crew respond to an emergency such as might require activation of the spacecraft's ejection seats.

Operational procedures with the GLV were also improved, with pre-fill chilldown adding 13,000lb (5,896kg) more propellant than could be accommodated by the same-size Titan II ICBM tanks, this due to the increased density. Moreover, the inertial guidance system which was standard on the ICBM version was replaced with a radio-guidance system for the GLV, operating from ground commands, and a general improvement in reliability raised the success rate while some redundancy in systems ensured there were fewer malfunctions resulting in an abort.

Assembly of GLV-1, for the first of two unmanned tests of both the launch vehicle and the spacecraft, began during September and October of 1962 with the Titan II contractor, the Martin Co, at its Baltimore, Maryland, facility. GLV-1 was delivered to Cape Canaveral on 26 October 1963 and erected at LC-19 three days later. The spacecraft had been manufactured by the McDonnell Aircraft Corp from St Louis, Missouri, and was delivered to the Cape on 4 October 1963. After several technical holds and postponements, not at all unexpected with a new spacecraft, a considerably modified launch vehicle and integrated tests were accomplished for the first time.

The launch of Gemini I occurred late morning on 8 April 1964, delivering the unmanned spacecraft to orbit still attached to the second stage of the GLV where it would remain for the

duration of the mission. The test programme required the spacecraft to be evaluated through telemetry for three full orbits of the Earth and it was not expected to survive re-entry – holes had been drilled into the heat shield to prevent that component of the spacecraft from surviving the searing heat of the atmosphere.

Because the aft section of the spacecraft was attached to the second stage it had no retro-rockets to fire and was expected to decay naturally out of orbit, the initial parameters being 204 miles (328km) by 99.6 miles (160.25km). But the orbit was 21 miles (34km) higher than planned, as a consequence of which the spacecraft and the attached second stage

ABOVE Gemini V being erected on LC-19 during its function serving as the launch pad for NASA's two-man Gemini spacecraft, the erector being in mid-position on its way back down to horizontal. *(NASA)*

LEFT Lift-off for the first Gemini-Titan from LC-19, the complex used for all 12 Gemini flights until it was retired in 1967 and completely erased by 2013. *(USAF)*

remained orbiting the Earth until 12 April. Post-flight results were good and the launch vehicle and spacecraft were qualified for a second unmanned test, this time of the all-important heat shield, together with astronaut simulator pallets to provide data on the habitability of the pressurised cabin and active retro-rockets to evaluate the overall performance of both launch and re-entry/recovery systems, a vital precursor to manned missions.

Manufacture of GLV-2 began in September 1962 and was completed by January 1964 with delivery to the Cape on 11 July 1964. Erected on 16 July, tests began the following day but a lightning strike on 17 August rendered those irrelevant due to possible damage. Eleven days later, Hurricane Cleo forced NASA to remove the second stage from the first and it was re-erected on 31 August. Then Hurricane Dora ripped through the Cape area and the entire vehicle had to be removed from the pad and re-erected on 14 September.

The spacecraft was shipped to the Cape on 21 September 1964 and mated with the launch vehicle on 5 November following which a succession of ground tests readied the stack for launch, which was planned for 9 December. Ignition occurred as planned but the engines suffered a sudden loss of hydraulic pressure and, sensing a malfunction, the stage shut itself down about 1.7 seconds later. The attempt was scrubbed and re-scheduled for 19 January 1965. NASA had hoped to get a manned flight off during 1964 to prevent a year-long gap since the flight of the last Mercury mission in 1963 but that had not been possible.

Launch occurred as planned and the Gemini II spacecraft separated from the spent upper stage just 5min 53sec later, reaching a maximum altitude of 98.9 miles (160km) and beginning a ballistic descent down toward the South Atlantic. Unlike the flight of Gemini I, Gemini II put the spacecraft through all the operational procedures for re-entry, including separation of the aft equipment section, retro-fire, re-entry and deployment of the parachute systems. All worked within tolerances and brought the spacecraft to a splashdown 18min 16sec after lift-off, 2,143 miles (3,448km) downrange, recovered by the USS *Lake Champlain*.

Cleared for manned flight, GLV-3 launched

the Gemini III spacecraft on 23 March 1965 with Virgil I "Gus" Grissom as command pilot and John Young as pilot. In all, LC-19 saw the launch of ten manned Gemini missions before the last on 11 November 1966, concluding a total of 27 Titan I and Titan II flights. At this date it was expected that the Air Force would utilise additional Gemini spacecraft for its Manned Orbital Laboratory (MOL) programme which would have been launched on Titan IIIC launchers from LC-40 but that was cancelled in 1969.

LC-19 was deactivated on 10 April 1967 and the service tower and umbilical tower were demolished on 30 May 1977. The white room, the enclosure from which the crew entered the spacecraft at the top of its launcher, was saved for posterity and in September 2003 it was restored and moved to the Air Force Space and Missile Museum. What remained of LC-19 was physically erased in 2012 and 2013.

LC-25B (14 August 1959) 28.4303°N x 80.5753°W

Matched with LC-25A (which see) but completed with access stand in January 1958. On 1 March 1957 a contract had been issued for a Polaris "ship motion simulator" which was designed to evaluate the design requirements for a missile launched from a submerged submarine in varying sea conditions imparting pitch and roll movement. This was installed on LC-25B and was itself first used on this initial date of operational use.

By 2 August 1960, when the last use of LC-25B was completed, a total of eight Polaris including one test vehicle and seven A1 types had been fired from this pad. The ship motion simulator was mothballed in October 1961 and the pad itself was dismantled in September 1969.

LC-29A (21 September 1959) 28.4296°N x 80.5771°W

Through a contract awarded on 12 August 1958, LC-29A was ordered to construction which was fully completed on 15 August 1959. Although it was built with a view to a second partner pad (LC-29B), that became the sole pad at this location and was formally accepted by the US Navy as a test site from were Polaris SLBM rounds could be fired down the Atlantic Missile Range.

Fourteen Polaris A1X test shots were fired between 21 September 1959 and 29 April 1960, followed by 15 Polaris A-2 from 10 January 1961 to 12 November 1965 and 18 Polaris A-3 from 7 August 1962 to 1 November 1967. There followed a series of three Polaris A-3 Antelope tests, a version of Polaris specifically developed for the UK's Royal Navy which had evolved from the US Navy's KH793 programme. It was developed in response to the construction of anti-ballistic missile (ABM) defences around Moscow.

The three Antelope tests took place between 17 November 1966 and 2 March 1967 and were part of a secret UK programme to retain the Navy's ability to decapitate the Soviet leadership by destroying the seat of government in Moscow. This had always been a British objective and an agreed role in association with the US's Single Integrated Operational Plan (SIOP). With ABM deployments around Moscow that ability was not so certain and could render the deterrent impotent if the Russians believed that too.

The story of Britain's nuclear deterrent is told in another Haynes Manual (*Nuclear Weapons* by this author) but suffice it to note here that the Antelope and Super Antelope anti-ABM programmes eventually resulted in the deployment of the Chevaline system which provided the UK with multiple independently targeted re-entry vehicles (MIRVs), bus platforms that could manoeuvre to evade ABM systems.

In addition to underground tests of the nuclear devices carried by Chevaline, there was a sustained programme of verification and training launches carried out by British submarines. At LC-29A, the pad had been placed on standby but it was reactivated in support of the Chevaline programme, with ten launches conducted there between 11 September 1977 and 19 May 1980.

LC-43 (25 September 1959) 28.4667°N x 80.5585°W

During the late 1950s a decision was taken to considerably expand facilities for launching sounding rockets and five pads were built (A–E), three clustered close together and two (D and E) a little way to the south. As the missile test programme shifted toward an emphasis

LEFT A Navy Polaris A3 test missile sits on LC-29A ready for flight in a series that kept this pad busy, on and off, for more than 20 years. *(USN)*

BELOW Not all launch pads supported large missiles and satellite launchers. LC-43 hosted scientific research flights with ballistic sounding rockets, such as the Loki with Rockoon seen here posed with Dr James A. Van Allen, who had instrumented Explorer 1. *(NASA)*

on the launch of high-value satellites and spacecraft across the Cape and the Kennedy Space Center, there was an increased need for weather data and much of this came from aerostat balloons and from the aeronomy and meteorological sounding rockets.

The need was particularly acute during the Shuttle programme and after a lightning strike on Apollo 12 in November 1969 there was increased attention paid to electrical storms in the vicinity. Moreover, weather information became important for Shuttle landing operations and these sites supported the need for that information also.

By 27 February 1984 about 4,680 rockets had been launched including a wide variety of Nike variants, Cajun Dart, Loki, Super Loki, Arcas and Hugo and many of these had been for general meteorological research outside the requirements for detailed weather information about the atmospheric column above the area.

During 1984 these weather operations were moved across to LC-47 in order to make room for construction of the new LC-46 to support Titan launches.

LC-16 (12 December 1959) 28.5017°N x 80.5518°W

One of four Titan pads built for America's second ICBM programme (see LC-15 on page 151), the contract to build LC-16 was awarded on 30 January 1957 and completed by July

1958. The Air Force accepted the site on 19 February 1959 ready for the first Titan I flight later that year, a launch that failed when a fault in the auto-destruct system detonated at ignition and blew the missile apart on the pad. The next two were also failures but for different reasons, the first successful Titan I flight from this pad occurring on 8 April 1960. Two more Titan I launches took place, both successful, the last on 27 May 1960.

Following modifications for the Titan II and a hiatus of almost two years, the first launch with this missile occurred on 16 March 1962, a re-entry test shot with the Mk 6 re-entry vehicle. A total of seven Titan II launches took place, the last on 29 May 1963, of which three were total failures. With the Titan II development flight series over, LC-16 was assigned to NASA in support of the Gemini programme where Gemini crew processing took place.

After the end of the Gemini programme in late 1966 it was converted into a static test facility for the Apollo Service Module's Service Propulsion System, its main engine designed to decelerate the docked Apollo and Lunar Module into lunar orbit and return the Apollo spacecraft to Earth. It was also responsible for conducting trajectory corrections between the Earth and Moon and for making lunar orbit changes with or without the Lunar Module.

The complex was deactivated and NASA departed in 1969 but three years later it was returned to the Air Force. In an inter-service shift, the US Army took it over during 1973 to prepare it for an extensive series of launch tests with the Pershing IA theatre missile, 88 being launched between 7 May 1974 and 13 October 1983. The site had been kept busy, up to six mobile launchers being positioned in unpaved areas directly in front of the blockhouse built originally for the Titan ICBMs.

LC-16 got a fourth career supporting the development of the much improved and more capable Pershing II which, along with the Ground Launched Cruise Missile (GLCM) was being deployed to NATO countries in Western Europe in direct response to the Soviet deployment of SS-20 ballistic missiles. The first Pershing II was launched from LC-16 on 22 July 1982, with the last of 49 on 21 March 1988. The test programme

BELOW LC-16 provides a backdrop to this dramatic launch of a Titan I on 8 March 1960, a site adapted a decade later for tests with the Pershing series of battlefield missiles. *(USAF)*

came to an end when Pershing II was deactivated as a weapon in accordance with the Intermediate Nuclear Force (INF) treaty which had been signed by President Ronald Reagan and Premier Mikhail Gorbachev on 8 December 1987.

In 2016 the private development company Moon Express moved on to the site to carry out trials with its tiny lunar lander.

LC-30/-30A (25 February 1960)
28.4391°N x 80.5792°W

Built exclusively for testing the Pershing I battlefield missile, construction began in December 1958 and was completed on 22 January 1960, accepted by the Army on 10 February. The complex included a dual pad configuration with associated service towers and facilities buildings for checkout and serviced by a helicopter pad. The blockhouse had an 8in (20cm) thick roof.

Launches began with flights from LC-30A on 25 February 1960, the last of 45 being launched on 21 March 1963. A total of ten Pershing I launches took place from LC-30 between 5 January 1961 and 24 April 1963. After the last launch from LC-30A the complex was decommissioned and the mobile gantry dismantled on 14 February 1968 and sold for scrap. In March 1968 the Navy took over the site and used it for assembling torpedoes which it employed for offshore sea tests.

LC-20 (1 July 1960)
28.5121°N x 80.5567°W

The last of the four Titan I facilities to start launch operations, LC-20 joined LC-15, LC-16 and LC-19. Following contract award on 23 April 1956, construction started in February 1957, completion logged on 10 September 1959 and acceptance the same day. Like so many launches at new pads, for inexplicable

ABOVE A Pershing II of the type used for extensive testing from LC-16 during the 1980s when NATO urgently sought an answer to the deployment of Russia's SS-20 theatre missiles. *(US Army)*

FAR LEFT With a short operational diary lasting a mere three years, LC-30/30A was built specifically for testing the Pershing I, seen here on 25 January 1961. *(US Army)*

LEFT The last of four pads built to support Titan I flight tests, LC-20 was used for the first Titan IIIA, seen here launching on 1 July 1960. *(USAF)*

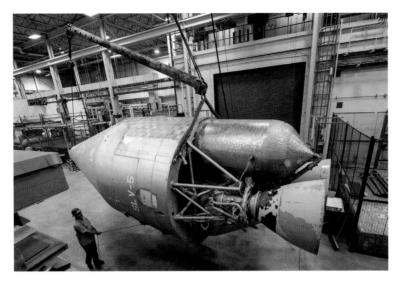

vehicle for the Gemini spacecraft and a new upper stage – Transtage – was developed to lift payloads of up to 6,800lb (3,100kg) to low Earth orbit and to propel military payloads to geosynchronous transfer orbit.

Transtage had been in development since August 1962 and would constitute the third stage of a modified Titan II, transforming it into the Titan IIIA. Transtage would do the job of an Agena, by restarting several times in space and manoeuvring so that it could deploy various satellites in different orbits. The basic Titan II had potential for launching much heavier payloads into space than it had been required to do for the Gemini programme, or for the Transtage three-stage configuration. Expansion was possible by adding solid propellant boosters either side of the core first stage and this configuration defined the Titan III, with or without additional boosters.

reasons the first was a failure, as was the second but the third launch on 30 August 1960 sent the Mk 4 re-entry vehicle to a maximum height of 600 miles (966km). The first "blue suit" launch with a wholly Air Force crew took place on 21 November 1961 and the last of 16 Titan I launches took place on 13 December 1961.

Like many pads of its type, LC-20 had to adapt to changing times and the need to support a new generation of rockets and launch vehicles. When Titan I flights were completed at LC-20 a period of adaptation made possible the launch of the Titan IIIA satellite launcher. Titan II had already been selected as the launch

The first Titan IIIA of the series took place from LC-20 on 1 September 1964 but the Transtage failed to pressurise the propellant tanks for taking over from the first two stages and it failed to achieve its objective when it cut off prematurely and failed to reach orbit. The second launch on 10 December was a success and the flight qualified the launcher for sending LES-1 into orbit for testing satellite communications techniques which could be of military application on 11 February 1965. The fourth and final Titan IIIA was launched on 6 May 1965 and that carried LES-2 and LCS-1, a large aluminium sphere used for calibrating radar. This success negated the value of a fifth Titan IIIA flight and the Transtage was considered qualified for more important missions with the Titan IIIC (see LC-40 for the first flight on 18 June 1965).

LC-20 had been destined to play a vital role in the Air Force Dyna-Soar boost-glide, trans-atmospheric vehicle which had been in development before the Space Age but that programme was cancelled in 1963. LC-20 was deactivated in April 1967 and the removable metalwork sold for scrap from 13 June that year. On 1 November 1988, however, work began at the site by the Butler Construction Company to build two new pads, A and B, supporting the Strategic Defense Initiative (SDI) Organisation's requirement for two Starbird launch pads.

Starbird was the ground-based element of a NASA Shuttle mission carrying out experiments for the Department of Defense in its Star Wars programme of ballistic missile defence, announced by President Reagan in 1983. Starbird rockets were to have been launched from LC-20 and from a facility at Wake Island in the North Pacific. A four-stage solid propellant suborbital rocket built by Orbital Sciences, Starbird would have appeared like a ballistic missile en route to its target.

The Shuttle would have been equipped with a European Spacelab in which was carried the Starlab suite of equipment to acquire and track the Starbird rockets as they ascended through the atmosphere, demonstrating the ability of a space-based system to successfully mark missiles in flight for destruction via some other offensive elements of the SDI programme. The mission was set for the STS-50 slot and assigned a launch date in May 1992 but the project was cancelled in 1990.

Considerable progress had been made adapting the site for Starbird, work being completed in December 1989 when it was turned over to the range. Each of the two pads supported 58ft (17.7m) long rails, separate launch equipment buildings and a payload assembly facility. The first Starbird test launch got off on 18 December 1990, the only such rocket in what was once considered to be the cornerstone of America's ballistic missile defence screen.

LC-20 did see continued use, however, with the launch of a Prospector sounding rocket, which failed on 18 June 1991, and an Aries launch vehicle converted from a Minuteman missile on 20 August 1991, another failure. Other test rockets followed with the last on 13 December 2000. The site is presently utilised as NASA's Advanced Technology Development Center (ATDC) which conducts work on new and innovative spaceport technologies and presently shares the site with the Florida Air National Guard.

LC-31A/B (1 February 1961)
28.4516°N x 80.5563°W

The first to be used in a unified complex, LC-31A was one of four pads built for development of the Minuteman ICBM at two launch complexes, seeing launches for all three variants of this solid propellant missile. The construction contract was awarded on 29 July 1959 with the essential brick structure completed in November 1960.

LC-31 was composed of pads A and B and a similar duo was established at LC-32 (which see). At each facility, pads A were built as conventional launch stands, the B pads being protective silos of the type the missiles would be deployed in at dispersed air bases. Providing launch services for both silo and pad, a single blockhouse was built with two storeys much like the Titan II blockhouses. The service tower was on rails and stood 65ft (19.8m) tall and weighed 190 tonnes. The silo was 90ft (27.4m) deep and 12ft (3.65m) in diameter with facilities supported by a large equipment room measuring 61.7ft (18.8m) by 17ft (5.18m) feeding under the adjacent pad.

A series of six Minuteman IA were launched from LC-31A by 20 February 1963 with a further series of Minuteman IA from LC-31B between 18 December 1961 and 19 November 1962. Fifteen Minuteman IB were launched from LC-31B between 14 December 1962 and 29 September 1964, after which LC-31 was modified for launches with the Minuteman II.

The first of this variant was launched from LC-31B on 6 December 1964, the last on 1 April 1966, after which the pad was placed in

BELOW Along with LC-31, LC-32 was dismantled after Minuteman I development and utilised for preserving wreckage from the Shuttle *Challenger* incident, which had occurred on 28 January 1986. *(NASA)*

ABOVE Home of the Saturn I ballistic flights prior to the introduction of the definitive two-stage launch vehicles (SA-5-SA-10), LC-34 was the first major construction at Cape Canaveral for NASA's ambitious programmes inherited from the von Braun team at the Marshall Space Flight Center. Clearly visible is the lone launch stand missing the umbilical tower yet to be installed, the large service structure which could be run up along a rail track and the dome-shaped blockhouse. *(NASA)*

temporary storage awaiting modifications to support test flights with the much improved Minuteman III, the variant still in service today. Four of this type were launched from silo LC-31B between 26 March 1969 and 23 September 1969, with deactivation of that pad immediately afterwards. However, LC-31A was used for a Minuteman II Mobile Feasibility test on 14 March 1970. In 1972 the Army adapted LC-31A for tests with the Pershing IA battlefield missile with 12 of this type being launched between 21 February and 19 March 1973.

After the last flights from LC-31, the complex went through the usual formalities of deactivation and closure. After the *Challenger* disaster of 28 January 1986, when the remains of the Orbiter had been forensically analysed and all information that could be gleaned had been extracted and documented, the remains of the Orbiter were interred within the LC-31B and LC-32B silos and capped off.

LC-32A/B (30 August 1961) 28.4537°N x 80.5556°W

This second launch complex paired with LC-31A/B was also built to support the development, test and qualification programme for the Minuteman missile system. The overall configuration as described for LC-31A/B was replicated here. LC-32 was unique in that, although built, no launches took place from the conventional launch hardstand at pad A.

Initially, there had been a mobile rail system built near LC-32A with 1,700ft (518m) of track to simulate the Mobile Basing System (MBS)

which was an early concept for Minuteman deployment. This was thought to be a means by which the actual launch location for the ICBM could be continuously moving so as to confuse Soviet targeting of counterforce missiles aimed at knocking out the US ICBM force. In December 1961 the idea was dropped in favour of deploying Minuteman in hardened silos.

Nine Minuteman IA missiles were launched from LC-32/B although the first got nowhere after it exploded the instant after ignition and two of the remaining eight were failures too. The last Minuteman IA from LC-32/B was launched on 9 August 1962. A total of 15 Minuteman IB were launched from silo LC-32/B between 7 December 1962 and 30 March 1964 before it was reconfigured for Minuteman II from late in that year. A total of 14 Minuteman II were launched between 24 September 1964 and 7 February 1968, after which 13 Minuteman III were sent up between 16 August 1968 and 14 December 1970.

See LC-31A/B (previous entry) for information on the use of LC-32/B for preserving the remains of the Shuttle *Challenger*.

LC-34 (27 October 1961) 28.5218°N x 80.5611°W

One of the most historic launch sites at Cape Canaveral, LC-34 was the first pad built for a space launch vehicle – Saturn I and IB – rather than a site for testing ballistic missiles. It was the pad where three astronauts (Grissom, White and Chaffee) lost their lives in a flash fire while rehearsing for their own manned flight. But it was to be the departure point for the first manned Apollo spacecraft into orbit – the last time a manned crew flew from Cape Canaveral Air Force Station.

Selection of the location for LC-34 took account of the power generated by the first stage of the Saturn I and by the energy yield of the stack should the vehicle explode on the pad or run amok during ascent. The Saturn I is described elsewhere and suffice to remind here that it was the product of the Army Ballistic Missile Agency through its Redstone Arsenal, later NASA's George C. Marshall Space Flight Center, Huntsville, Alabama, in response to a requirement issued in 1956 for a very large launch vehicle to place in orbit some heavy military satellites then being considered for communications purposes.

The general requirement was for a launcher capable of placing up to 40,000lb (18,000kg) in orbit or sending 11,900lb (5,400kg) to escape velocity. Calculations showed that this would require a launch vehicle with a first stage thrust of at least 1.5million lb (6,700kN) which was 3.5 times the thrust of America's largest rocket at the time – the Titan I, albeit still in development and three years away from its first flight. The decision was made to cluster existing rocket motors and assemble tanks from existing missile production lines for the Jupiter and military rockets which had been developed at Redstone Arsenal.

The Saturn rocket had no military value as a weapon but rather as a heavy-lifter for ambitious space missions. When NASA was formed in 1958 it was by no means certain that all space programmes would coalesce around the new civilian agency and both the Air Force and the Army continued pushing for their own uniquely slanted requirements met through their own projects and programmes. Thus it was that even when consideration was being given to merging the ABMA's Saturn programme with NASA the Army still considered its own application of Saturn I in the running.

The selection of a launch site for this giant rocket initially examined a variety of different locales and options for how to launch such a large rocket were studied in detail, including launching from an offshore caisson, in this way minimising the acoustic and shock effect and providing a safe location should anything go wrong. Eventually, a conventional launch site with hardstand was decided upon and a construction award granted for LC-34 on 3 June 1959. The pace was commensurate with an anticipated first launch in October 1960. At the time the Saturn I was known as Juno V, a progression from the Juno II launcher, but it would assume the more popular name when the project was transferred to NASA.

Construction of LC-34 began five days after contract award and work proceeded rapidly, final inspection of the completed blockhouse being accomplished on 11 August 1960. The overall design of the complex was based on the principle of vertical assembly of stages and elements on the pad itself. It included a fixed service mast adjacent to the assembled stack, a mobile service tower for working on the stack pre-launch, a blockhouse for controlling events and supporting infrastructure including propellant tanks and hydraulic, pneumatic and power supplies.

BELOW LEFT Saturn SA-3 prior to launch, with the large service tower in place and the umbilical tower just behind the rocket, carrying dummy upper stages. *(NASA)*

BELOW SA-3 lifts off on the third flight of a Saturn I and the penultimate single-stage Saturn launch from this pad before it was converted for the Saturn IB. *(NASA)*

ABOVE President Kennedy visited the LC-34 blockhouse on 16 November 1963 and received a briefing from Dr George Mueller (left out of frame), with (left to right) NASA Administrator James Webb, Vice President Lyndon Johnson, KSC Director Dr Kurt Debus and President Kennedy. Secretary of Defence Robert S. McNamara is second from right, hand on chin. (NASA)

LC-34 was built on 45 acres (18.2ha) of land supporting a central pad built up from reinforced concrete, 438ft (133.5m) in diameter and 8in (20.3cm) thick. It incorporated a central pedestal 42ft (12.8m) square and 27ft (8.2m) tall with a foundation containing 4,400yds^3 (3,364m^3) of concrete and 580 tons (526 tonnes) of steel varying in depth from 8ft (2.4m) at the centre to 4ft (1.2m) at the edges. Bolted to a ring at the top of the pedestal were eight arms, four for supporting the rocket and four for both support and restraint from ignition to lift-off when all eight engines were running correctly.

A steel structure situated beneath the pedestal consisted of a rail-mounted flame deflector which would divert the 5,000°F (2,760°C) flame from a 26ft (7.9m) square opening to two opposing exits. The deflector was 43ft (13m) long, 32ft (9.7m) wide and 21ft (6.4m) tall, weighing 150 tons (136 tonnes). A spare deflector was parked on a spur track on the same side of the pedestal. An umbilical tower was fixed adjacent to the pedestal. For the first launch it was only 27ft (8.2m) tall but later when upper stages were attached the tower had a total height of 240ft (73m) and was 24ft (7.3m) square.

The service structure consisted of two vertical truss legs connected by an arching structure at the top. It was 310ft (94.5m) tall with each leg of the tower 70ft (21.3m) by 37ft (11.27m) at the base. The space between the two legs was 56ft (17m) to accommodate the Saturn rocket when the service structure was wheeled in on rails to enclose the stack for checkout. A bridge crane with a 54.4 tonne capacity was located centrally atop the structure to manoeuvre the stages into position.

At the time of construction this was the largest moving structure in the world and was supported on two standard-gauge rail tracks. A single operator could control the movement of the structure via a cab at the 27ft (8.2m) level and after checkout it would be moved back

LEFT President Kennedy gets a tour of LC-34 with von Braun (centre) and Associate Administrator Robert C. Seamans (left) in front of a model of a Saturn I, by which date the last Saturn I to fly from LC-34 had been launched. (NASA)

RIGHT After some modifications, flight operations were scheduled to return to LC-34 in February 1967 for the first flight of Apollo 1, the initial flight of a crewed spacecraft in this series. This crew access arm allowed ingress to the Command Module. *(NASA)*

a distance of 600ft (182.9m) from the launch pedestal. It was moved thus for the first time on 26 March 1961.

The launch centre, or blockhouse, was based on the pattern for Atlas and Titan at other pads at the Cape. Located 1,000ft (305m) from the launch pedestal, the blockhouse had a diameter of 120ft (36.5m), was 30ft (9.1m) high and 5–7ft (1.5–2.1m) thick. It had 10,000ft^2 (929m^2) of protected floor space on two levels and an additional 2,150ft^2 (199.7m^2) of unprotected space in an adjacent equipment room which was not occupied during launchings. Final inspection of the blockhouse took place on 11 August 1960.

The first floor of the building was used for the first and upper stage contractor personnel involved in tracking and telemetry operations. The main firing room was located on the second floor with consoles, test supervisors, conductor consoles and several monitoring and recording equipment stations. A small observation room was separated by glass from the operating area with pre-launch activities in the area viewed from an observation balcony on top of the blockhouse.

Propellant was supplied to the vehicle from two LOX and one RP-1 container, spherical vessels which supplied the rocket with liquid oxygen and kerosene in the pre-launch countdown. LOX was contained in two storage tanks, each with a capacity for 125,000 US gallons (473,125 litres), situated 650ft (198m) from the pedestal with a 6in (15.2cm) diameter supply line with a flow rate of up to 2,500 US gallons/min (9,462 litres/min). It took 40 minutes to fill the Saturn I with liquid oxygen. The two RP-1 tanks had the same capacity as the two LOX tanks and were located 950ft (289.5m) from the rocket and, uniquely, was fully automated on command from the control building, also taking 40 minutes to completely fill the fuel tank of the Saturn I. When the Saturn IB derivative was

LEFT After the Apollo 1 fire of 27 January 1967, LC-34 hosted the launch of Apollo 7 on 11 October 1968, the first manned Apollo mission carrying astronauts, Schirra, Eisele and Cunningham, after which it was deactivated. *(NASA)*

BELOW An evocative shot of fire and steel as AS-205 lifts off from LC-34, the last flight from this complex. *(NASA)*

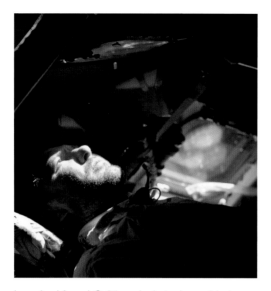

RIGHT From the Commander's couch, Walter M. "Wally" Schirra gazes through the left forward-facing window, the side window to his left, in a rare moment of contemplation during Apollo 7's mission in Earth orbit. (NASA)

BELOW LC-34 remains a commemoration to Grissom, White and Chaffee, lives lost on that day when the harsh realities of ever-present danger were tragically realised. (NASA)

INSET A plaque in remembrance of the lost crew of Apollo 1. (Via David Baker)

launched from LC-34, a single tank provided 125,000 US gallons (473,125 litres).

High pressure gas facilities provided the nitrogen and helium which were essential for preparing the launch vehicle for flight, located about 1,100ft (335m) from the pedestal, consisting of 36 storage vessels in two groups. Four vessels contained the helium used for bubbling the LOX tanks on the rocket while 32 contained the nitrogen for purging the oxidiser and fuel lines, engine and instrument compartments and supported air bearings and pressure-operated components such as valves.

A skimming basin was located 300ft (91m) on the shore side of the complex, a holding vat used for collecting fuel which may have spilled

on the pad and which was thus prevented from entering the drainage channels. Water systems were installed at the pad and throughout the service structure as both a safety measure for fire protection and as a quenching system for any fire which may break out in the boattail area, or to extinguish fire which may break out in the engine compartment at the base of the first stage should the stage have to be shut down on the pad.

Occupation of LC-34 occurred beginning February 1961 as preparations were made for the first flight of the Saturn I. On this first flight (SA-1) only the first stage was live but the rocket performed as planned, reaching a height of 85 miles (137km) and a downrange distance of 215 miles (346km). SA-2 followed on 25 April 1962 and concluded by splitting open the simulated upper stage to release 22,900 US gallons (86,676 litres) of water ballast into the upper atmosphere at a height of 65 miles (105km) for scientific analysis of the effect. SA-3 was launched on 16 November 1962 and repeated the water experiment followed by SA-4 on 28 March 1963 with a successful demonstration of an engine-out capability whereby the remaining seven H-1 engines continued to burn longer to compensate for the thrust decrement.

For nearly three years Saturn operations switched to LC-37 where it matured by adding the S-IV upper stage which allowed the rocket

to deliver payloads to orbit for the first time. A further six Saturn IB launchers flew from that facility while modifications were made to LC-34 for flights with the more powerful S-IVB upper stage. With this, and a more powerful first stage, the launcher became the Saturn IB, the first flight of which (AS-201) took place from LC-34 on 26 February 1966 delivering an Apollo Block I spacecraft to a suborbital trajectory for the first time, reaching a maximum altitude of 306 miles (492km) and a downrange distance of 5,268 miles (8,477km).

The second Saturn IB flight (AS-203) took place from LC-37 but the third (AS-202) was launched from LC-34 on 25 August 1966 on another test of the Block I Apollo spacecraft on a trajectory designed to test its heat shield and for launcher qualification. The AS-204 Saturn IB was being prepared for the first manned Apollo flight scheduled for late February 1967 when a flash fire during a countdown simulation at LC-34 on 27 January took the lives of astronauts Grissom, White and Chafee.

The AS-204 launcher was moved to LC-37 for the unmanned flight of the first Lunar Module while AS-205 was used to launch Apollo 7 on 11 October 1968 from LC-34. Crewed by astronauts Schirra, Cunningham and Eisele, the flight lasted almost 11 days, demonstrating the technical integrity of the spacecraft and qualifying it for a flight around the Moon with Apollo 8 from LC-39A two months later. Those Apollo 7 astronauts would never fly in space again but what they achieved returned confidence that the Apollo programme could achieve its goal of landing astronauts on the Moon by the end of the decade.

Originally, a second pad (LC-34B) was planned to the south but this was cancelled when it was realised that this would place it only 2,400ft (730m) from LC-20 and this resulted in construction of LC-37 (which see). When Saturn I flights began it was expected that this launcher would carry early Apollo crews on Earth-orbit shakedown flights but that was considered unnecessary and only the one Saturn IB crewed Apollo mission was sufficient to shift missions to the massive Saturn V.

On 29 November 1968, following the flight of Apollo 7, a memorandum was issued for the deactivation of LC-34, as well as LC-37, thus ending a historic period in the history of Cape Canaveral and the Kennedy Space Center, focusing all manned space flight thereafter on LC-39.

LC-36A (8 May 1962)
28.5336°N x 80.5378°W

This launch complex was built specifically to accommodate the Atlas-Centaur launch vehicle, an evolution of the Atlas ICBM married to the cryogenic Centaur upper stage using high-energy LOX/LH2 propellants for greater efficiency in payload lifting capacity. Centaur itself had been a troubled development, begun first by the Marshall Space Flight Center which had pioneered the development of the S-IV stage for the Saturn I and then been transferred, amid much objection, to the Lewis Research Center.

Atlas-Centaur was pivotal to the success of so many space missions that this launch complex must go down as one of the most historic sites at the Cape and this summary cannot do full justice to the outstanding number of satellites and spacecraft sent on their way from this facility.

Built initially as a single pad, LC-36A was completed in February 1961 but bids for a second pad were issued later that year. The first launch attempt on 8 May 1962 was a failure when the Centaur exploded 55 seconds after lift-off but the second, on 27 November 1963, was a success. Built by the Air Force,

BELOW LC-36A prepares to send Pioneer 10 on its way to Jupiter, a launch which occurred on 3 March 1972. *(NASA)*

management shifted to NASA in 1963. The third launch, on 30 June 1964, which was to have placed a payload in geosynchronous orbit was also a failure. But the third launch, on 11 December, was a success but this was followed on 2 March 1965 with the final development flight for what had been a troubled and lengthy gestation for this technologically advanced cryogenic stage. It too was a failure.

Atlas-Centaur would go on to become one of the most successful launch vehicles of all time and give the old ICBM an entirely new lease of life, this combination taking over from the Atlas-Agena of the early years. One programme in particular was entirely reliant on Atlas-Centaur, the Surveyor soft-landing Moon probe which was itself one of the most complex unmanned spacecraft launched to that date.

Against all odds, Surveyor succeeded on the first fully operational flight of this launcher on 30 May 1966, becoming the first spacecraft to make a controlled descent to the surface of another world on 2 June. Russia's Luna 9 had landed on the Moon on 3 February that year but that had been a survivable hard impact and not under the control of throttleable descent engines. Five of the total of seven Surveyor spacecraft were successful, the last launched on 7 January 1967, with three launched from LC-36B.

Other successes for Atlas-Centaur followed, with Mariner 7 launched on 27 March 1969, the first Mars fly-by since the successful Mariner 4 mission four years earlier. Applications technology satellites in the ATS programme followed, with communications satellites

following. Another "first" for LC-36A was the launch of Pioneer 10 on 3 March 1972, the first spacecraft to cross the asteroid belt and the first fly-by of Jupiter (in 1973). For the next several years this launcher carried communications satellites into orbit until the Pioneer Venus Orbiter mission of 20 May 1978 followed by the Venus Multiprobe mission launched on 8 August that year, another succession of communication satellites following.

On 1 December 1988 NASA awarded a contract for a new umbilical tower to allow the introduction of the modified Atlas II/Centaur launch vehicle at LC-36A and significant work was carried out on the existing Mobile Service Tower (MST) to stem corrosion, always a problem at the Cape. Work was completed in October 1989 and the first launcher of this type to fly from this pad lifted off on 11 February 1992, the last of nine Atlas II being launched on 16 March 1998. A total of 28 Atlas IIA/AS were flown between 3 August 1994 and 31 August 2004, and the last of 63 Atlas IIA/AS launchers lifted off LC-36A on 31 August 2004 carrying a classified payload for the National Reconnaissance Office.

This was also the last of 576 Atlas rockets powered by the original Rocketdyne rocket motors. Under the management of the USAF 45th Space Wing, in 2007 LC-36 was demolished, the Mobile Service Tower for LC-36A blown up on 16 June preceded by 12 minutes by the MST for LC-36B. The vacated site was used for Moon Express tests with its mini lunar lander and in 2015 Blue Origin took a long-term lease on the site for its future launch vehicles; Moon Express moved its tests across to LC-17 and LC-18 in 2016. Blue Origin broke ground in June 2016 for its New Glenn launch vehicle which, with a reusable first stage, will have a lift-off thrust of 3,850,000lb (17,125kN).

LC-37A/B (29 January 1964) 28.4713°N x 80.5680°W/ 28.5313°N x 80.5644°W

Planned as dual launch pads, LC-37A was never used for a launch but LC-37B saw the first launch of the Saturn I in orbital configuration and as such is a historic site in its own right. This was also the flight in which NASA pulled level with Russia in lift capacity, which the late President Kennedy had noted as the point at which he felt the nation had vindicated itself for Sputnik and the flight of Yuri Gagarin.

Bids to build LC-37 were invited in March 1961 with site preparations getting under way the following month and construction completed on LC-37B on 7 August 1963 followed by LC-37A ten days later. Each of the two pads had their own launch stand and umbilical tower with a common servicing tower moved between pads along a rail track. When built it was the largest moving structure in the world and jacks were available at each pad to take the weight off the wheels when in position.

Each pad was 300ft² (27.87m²) in area with the pedestals 55ft (16.7m) square and 35ft (10.7m) high incorporating eight identical hold-down arms. The flame deflectors were 43ft (13.1m) in length, 32.3ft (9.8m) wide and 21ft (6.47m) high, weighing 136 tonnes and were fabricated from ASTM A36 steel of dry, roof-truss structure. The umbilical towers had a height of 268ft (81.68m) tall with a base 32ft (9.75m) square and 35ft (10.67m) tall, tapering to 14ft (4.26m) square at the top. The two fixed towers for respective pads were 900ft (274m) apart and carried one swing arm for the first stage, two for the S-IV stage and one for the spacecraft.

Launch Pad 37A

Launch Pad 37B

The single service structure had a height of 300ft (91.4m) excluding the top derrick, with a base 120ft (36.57m) square with a total weight of 3,175 tonnes, a structure moving between pads as required along the 1,200ft (365.76m) rail. The structure was sized to accommodate the growth versions of Saturn envisaged in 1961 when the design specification was finalised.

RP-1 fuel was stored in an earth-revetted cylindrical tank with a length of 67ft (20.4m) and a diameter of 12ft (3.65m) with a capacity of 43,500 US gallons (164,647 litres). Liquid oxygen was stored in two spherical double-

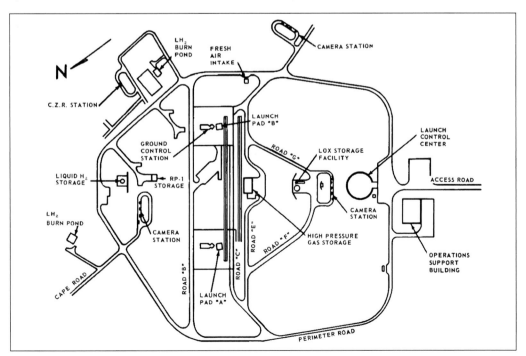

RIGHT A clear view of LC-37 with the common track for the one service structure, also shown, the flame bucket and SA-5 on the launch stand, the first to fly from this complex. *(NASA)*

BELOW A clear view of the rear of the LC-37 service structure as the SA-7 vehicle is prepared for flight. *(NASA)*

BELOW RIGHT AS-204 on LC-37 prior to the flight of LM-1, the first Lunar Module, on 22 January 1968, the last of only eight launches before the site was deactivated after the decision to launch Saturn IB rockets in support of the Apollo Applications Programme evaporated. *(NASA)*

walled tanks with an outside diameter of 42.5ft (12.87m) and a capacity of 125,000 US gallons (473,125 litres). Liquid hydrogen for the S-IV and S-IVB stages was contained in a double-walled spherical tank with a diameter of 38.8ft (11.8m) and a capacity of 125,000 US gallons.

The launch control centre consisted of a two-storey domed building located 1,200ft (365.76m) from the pad capable of surviving a blast pressure of 2,188lb/in² (15,000kPa). The inner shell was of 5ft (1.52m) thick reinforced concrete earth-revetted 7ft (2.1m) at the top and 41ft (12.5m) at the base. An internal layer of 2in (5cm) soundproofing material helped control the acoustic effect of launch. Close by, an operations

support building had a length of 101.9ft (31m) and a width of 40ft (12.19m) and a separate spare parts storage facility had a length of 162ft (49.37m) and a width of 42ft (12.8m).

On acceptance of the launch complex in August 1963 preparations began immediately for the first flight of the two-stage Saturn I, the first four launches of which had been ballistic flights from LC-34 with only a live first stage. The flight of SA-5 took place at 11.25am on 29 January 1964 and was a complete success, the first stage achieving full thrust of 1.5million lb (6,672kN) with the S-IV cryogenic upper stage providing 90,000lb (400.32kN) of thrust from its six RL-10 rocket motors. This was also the first test of the Saturn I's Block II guidance system.

SA-6 followed on 28 May 1964 with an Apollo boilerplate and launch escape tower (SA-5 had carried a Jupiter nose cone) and this too was a success, demonstrating launch vehicle/spacecraft compatibility. The last four Saturn I flights followed: SA-7 on 18 September 1964; SA-9 on 16 February 1965; SA-8 on 25 May 1965; and SA-10 on 30 July 1965. The last three flights carried large micrometeoroid detection satellites known as Pegasus, consisting of large, wing-like arrays deployed in orbit to record the number and size of particle impacts in low Earth orbit.

Following the first flight of a Saturn IB from

LC-34 on 26 February 1966 (which see), the second flight (AS-203) took place from LC-37 on 5 July 1966 as the first orbital test of this rocket, placing the S-IVB stage in orbit but with a 28ft (8.5m) aerodynamic fairing and no Apollo payload. The object of this flight was to test the low-gravity behaviour of the S-IVB as it was to be the third stage of the Saturn V and required to fire a second time to push the Apollo payload out of Earth orbit and on to a trans-lunar trajectory.

Delayed due to programme repercussions from the Apollo fire of 28 January 1967, and by delays to the development of the spacecraft, the first test of a Lunar Module began with the launch of the fourth Saturn IB (the second from LC-37) on 22 January 1968 when AS-204 placed LM-1 in low Earth orbit for a successful qualification of the Moon lander. After just eight launches, LC-37 was deactivated on 1 January 1969, subsequent launches of the Saturn IB taking place from LC-34 and from LC-39. The pad was mothballed in November 1971 and the service structure for LC-37B was scrapped in April 1972, the complex returned to the Air Force in November 1973.

Some 30 years after its last launch, Boeing received a right of entry on 8 January 1998 in support of plans to introduce its new Delta IV launch vehicle, a successful bidder in the Evolved Expendable Launch Vehicle (EELV) contract for

military payloads. This would require considerable redesign to the site to accommodate a new generation of very powerful launch vehicles. In August 1999 work began on the Horizontal Integration Facility (HIF) in which the launch vehicle would be prepared in a horizontal orientation and moved to the pad on a rail track where it would be raised to a vertical position ready for installation of the payload a few days prior to flight, final checkout and launch.

Under the reactivation of the site the designation changed to Space Launch Complex 37 (SLC-37) and included a single launch pad, a Mobile Service Tower (MST), a Common Support Building (CSB), a separate Support

ABOVE Completely redeveloped, LC-37 is now the home of the Horizontal Integration Facility for the massive Delta IV launch vehicle. *(NASA)*

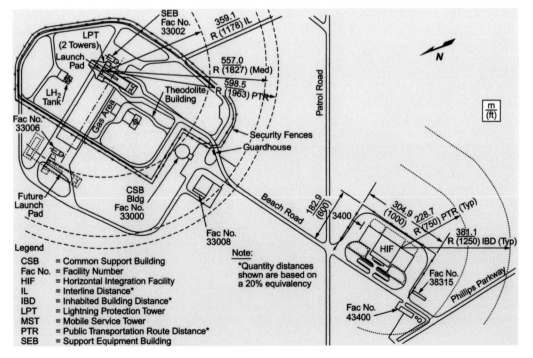

Legend

CSB	= Common Support Building
Fac No.	= Facility Number
HIF	= Horizontal Integration Facility
IL	= Interline Distance*
IBD	= Inhabited Building Distance*
LPT	= Lightning Protection Tower
MST	= Mobile Service Tower
PTR	= Public Transportation Route Distance*
SEB	= Support Equipment Building

Note:
*Quantity distances shown are based on a 20% equivalency

LEFT Reflective of the new age of heavy launchers demanded by commercial users, LC-37 has been completely transformed, as this layout plan reveals. *(ULA)*

Equipment Building (SEB) and associated support infrastructure. Adhering to the trend in launch vehicle design, where the first stage is supplemented by powerful booster rockets – usually of solid propellant – the HIF had a very different arrangement and configuration of buildings, the assembly of the vehicle in a horizontal orientation being one.

Moreover, the optional configuration of the launch vehicle depending on payload size and mass, required greater flexibility in the assembly stage. The smallest Delta IV consists of a Common Booster Core (CBC) which carries no boosters, referred to as a "single-stick" configuration, while the intermediate rockets have two or four small strap-on boosters derived from the Delta II and Delta III programmes. The largest Delta IV-Heavy consists of three CBC units strapped together in a triple-parallel configuration with a lift-off thrust of more than 2.1million lb (9,340kN). All this informed the design of SLC-37 and the buildings required.

The MST is 330ft (100.5m) tall, topped

out with a 45 tonne bridge crane with a 300ft (91.4m) hook height, and is moved to the service position using a hydraulic drive system with pneumatically and hydraulically driven platforms designed so as to accommodate the five different configurations of Delta IV and the variable dimensions of the upper stages and payload fairings. The HIF is not directly part of the SLC-37 pad area but is responsible for assembling the selected configuration. It consists of a seven-storey building incorporating two bays which can accommodate four single-stick rockets or two single-stick Deltas with strap-on boosters. Each bay is 250ft (76.2m) by 100ft (30.5m) and has a 22.67 tonne utility bridge crane and a 74ft (22.6m) door at each end.

When the rocket is delivered to the pad it is attached to the Fixed Pad Erector which rotates the vehicle from a horizontal to a vertical position and on to the launch table which measures 65ft (20m) wide by 45ft (13.7m) long and 23ft (7m) tall. The pad has a 250,000 US gallons (946,250 litre) liquid oxygen tank and an 850,000 US gallons (3.217million litre) liquid hydrogen tank.

The new lease of life for SLC-37 began with the launch of the first Delta IV on 20 November 2002 carrying the Eutelsat W5 satellite for the European telecommunications consortium, commencing a move away from the old style Delta II, although that launcher would continue to fly until 2017. Several classified launches took place after that but NASA returned on 24 May 2006 for the launch of the GOES-13 weather satellite. This was followed by a series of additional GOES launches interspersed with military satellites.

A milestone in NASA's bid to return to human space flight took place on 5 December 2014 when a Delta IV-Heavy launched an unmanned Orion spacecraft on Exploration Flight Test-1 (EFT-1), sending the spacecraft to a maximum altitude of 3,610 miles (5,809km) before it returned to Earth and a successful splashdown off Baja, California, some 4hr 24min after launch. SLC-37 continues to be used for Delta IV launches.

BELOW The Horizontal Integration Facility (HIF) allows for simultaneous preparation of launcher elements, a transformation from the conventional vertical stacking of the early days of the space programme. *(ULA)*

LC-40 (18 June 1965) 28.5620°N x 80.5772°W

With the decision in the early 1960s to use the Titan II ICBM as a satellite launcher, this rocket saw a wide range of upper stage configurations fulfilling that requirement. LC-40 was built to support flights with the Titan IIIC, using a dual pad arrangement involving this together with LC-41 connected via an Integrated Transfer Launch (ITL) facility, the two pads and the ITL being located at the north end of Cape Canaveral.

The ITL concept emerged from the Large Launch Vehicle Planning Group composed of staff from the Department of Defense and NASA. Recognising the need for a much bigger launch vehicle than existing types the Titan IIIC evolved to dramatically increase the lift capability by attaching two solid rocket boosters to the first stage, which was extended in length. With a lift-off thrust of 2.34million lb (20,408kN) the two solids burned for 120sec before falling away leaving the first stage to ignite followed by the Transtage upper stage in sequence.

The Titan IIIC provided the significant increase in payload capacity required but the need to accelerate the preparation time pushed development of the dual-pad/ITL concept which was as significant a development in several respects as NASA's LC-39 and Moonport facilities still farther to the north. To construct the ITL complex some 6.6million yds³ (6million m³) of fill was dredged from the Banana River to make three artificial islands to incorporate a 19.9 mile (32km) rail system, a storage area for the solid rocket booster segments, a Vertical Integration Building (VIB), Solid Motor Assembly Building (SMAB) and transporter system and the two launch complexes.

The core segments would be delivered to the SMAB and brought together in a single stack and integrated with the solid rocket motors in the VIB, together with any payloads. Thus assembled, the completed vehicle would be moved to either LC-40 or LC-41 for launch approximately five days later. On later missions, the cryogenic Centaur upper stage was used for delivering heavier loads to Earth orbit or sending interplanetary probes on their way.

LC-40 is historic for several reasons, not least that it hosted the inaugural flight of the Titan IIIC. However, while the second launch, on 15 October 1965, got away successfully and entered orbit, a restart of the Transtage motor fired to push the three satellites into their assigned orbits failed when the stage exploded.

The third launch on 3 November 1966

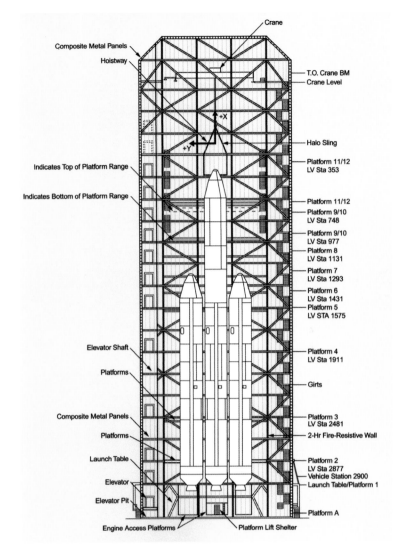

successfully carried a mock-up of the Air Force Manned Orbiting Laboratory (MOL) together with a re-launch of the Gemini II spacecraft which NASA had sent into space on a Titan II on 19 January 1965. The spacecraft was successfully recovered, the first time a manned space vehicle (albeit unmanned on these occasions) had been sent into space and successfully recovered twice. Several more military launches occurred before NASA made its first use of the ITL and LC-40.

On 30 May 1974 NASA launched its sixth Applications Technology Satellite (ATS-6) to a geosynchronous orbit from where it was used on several occasions to migrate to different longitudes including positions over the Americas, the Indian Ocean in 1976 and to a drifting trajectory from 1979. The Air Force introduced the much more powerful Titan 34D from 30

ABOVE The Mobile Service Tower for LC-37 allows for various payload configurations and servicing levels in a structure 330ft (100.5m) tall. *(ULA)*

LEFT With the decision to develop a heavy lift launcher from the Titan II series and add solid rocket boosters to produce the Titan IIIC, LC-40 emerged as an Integrated Transfer Launch (ITL) facility hooked up to LC-41, essentially a common launcher and payload preparation area for delivery to either complex. *(USAF)*

October 1982, supporting the launch of two military communication satellites with a solid propellant Inertial Upper Stage (IUS) for the final kick into orbit, or on some flights the Transtage.

A succession of launches saw a wide variety of military payloads placed in orbit and on 1 January 1990, the British Skynet 4A military communications satellite was successfully delivered to the first of several different longitudinal locations. This satellite had been assigned to a Shuttle launch until the *Challenger* disaster of 28 January 1986 resulted in the transfer of many satellites to expendable launch vehicles. Skynet 4A had been the first use of the Titan III (also known as Commercial Titan) and was followed by two successive launches of commercial communications satellites, followed in turn on 25 September 1992 when NASA again used LC-40 to launch its Mars Observer mission. The launch phase was a success but the probe failed during Mars orbit insertion.

This was the fourth and last flight of Titan III, which had been unsuccessfully marketed as a commercial venture by Martin Marietta (the Titan contractor) when the Air Force chose the Delta II over this derivative for its medium-lift expendable launcher competition. The definitive heavy-lifter from this family, the Titan IV had first flown from LC-41 (which see) on 14 June 1989 but was introduced at LC-40 with the launch of an advanced military communication satellite on 7 February 1994.

Another NASA mission departed LC-40 on 15 October 1997 when the Cassini/Huygens spacecraft left for Saturn where it arrived on 1 July 2004 after multiple fly-bys including Venus, Earth, the asteroid Masursky and Jupiter for gravity assist. This was the last NASA mission from LC-40 but more classified launches followed with the last of 17 Titan IV from this site on 30 April 2005. This was the beginning of the end for the Titan launcher programme and the site was placed on standby pending a decision about using LC-40 for Ares launches in NASA's Constellation programme.

The Mobile Service Tower was demolished on 27 April 2008 as the site was prepared for a new tenant – SpaceX with its Falcon 9 programme, changes which involved the erection of a large hangar for preparation of the rocket, a new launch stand, a mobile Transporter/Erector system, and various small support buildings. The first Falcon 9 test launch got off the modified pad on 4 June 2010 with the Dragon cargo capsule which was successfully recovered in the Pacific Ocean on the second orbit.

The next launch saw a Falcon 9 send the second Dragon capsule to orbit on 22 May 2012 which docked to the International Space Station three days later, the world's first commercial delivery of cargo to a space station. The second cargo delivery got off on 8 October 2012 but the Falcon 9 suffered malfunctions which did not, however, prevent successful delivery of the CRS-1 Dragon to the ISS. CRS-2 followed on the next launch on 1 March 2013 but further problems delayed docking by one day.

The next flight on 6 January 2014 sent a Thai communication satellite to geostationary orbit followed by another Dragon delivery to the ISS on 18 April 2014. A cluster of communication payloads went up on 14 July, followed by Asiasat 8 on 5 August and Asiasat 6 on 7 September. Dragon CRS-4 went to the ISS on 21 September with CRS-5 following on 10 January 2015. The relentless pace of Falcon 9 flights from LC-40 continued with the launch of Discovr on 11 February 2015 – a satellite originating with an idea from Vice President Al Gore to send a satellite into orbit with the capacity to continuously photograph the Earth as an eco-message to humanity.

The steady and consistent stream of Dragon flights interspersed with commercial launches on the Falcon 9 was broken only by a catastrophic accident on 1 September 2016 when a malfunction during preparations for a static fire test on the pad resulted in an explosion in the second stage, loaded with propellants to simulate the full mass of the stack, and the loss of the satellite. The site

returned to service with the launch of a Dragon cargo vehicle (CRS-13) on 15 December 2017 and continues in service for Falcon 9 launches to the ISS and for carrying commercial satellites into orbit.

LC-36B (11 August 1965) 28.4682°N x 80.5410°W

Essentially similar in layout to LC-36A, this site was constructed beginning in February 1963 and completed in July 1964. The first three

LEFT LC-40 supported the launch of the simulated MOL with Gemini II on a Titan IIIC sent up on 3 November 1966. *(USAF)*

BELOW LC-40 now hosts SpaceX for its Falcon 9 launches, a commercial leasing of a classic Cape Canaveral launch complex now that the ITL and the Titan families have been retired from use. *(SpaceX)*

flights were Atlas-Centaur test launches carrying dummy Surveyor payloads but the fourth launch, on 17 April 1967, carried the Surveyor 3 spacecraft which successfully landed on the Moon three days later. It was this spacecraft which was visited by Apollo 12 astronauts Conrad and Bean in November 1969 on the second manned lunar landing, with sections of the spacecraft cut off and returned to Earth.

Compared with its sister pad, LC-36B had a slow start to launch operations, with only one flight in 1969, 1970 and 1971. But the flights from this pad were highly significant. Mariner 6 left for its Mars fly-by on 25 February 1969 followed by the first Mars orbiter, Mariner 9, on 30 May 1971. From early 1973 the pace picked up with communications satellites making full use of the expanded lifting capacity of Atlas-Centaur to deliver payloads to geosynchronous orbit. On 6 April 1973 Pioneer 11 followed Pioneer 10 (launched from LC-36A) on a fly-by to Jupiter but this time on to pass Saturn too, in 1979.

Another historic planetary mission got off the pad on 3 November 1973 when Mariner 10 began a fly-by mission to Venus on its way to an encounter with Mercury in March 1974 – the first to visit the innermost planet – where it entered a synchronous loop around the Sun, visiting Mercury twice more with fly-bys in September 1974 and March 1975.

Between 9 June 1984 and 25 September 1989 eight Atlas G/Centaur were launched, all but the first, the launch of an international communications satellite, a success. This launcher was a stretched version of Atlas and with improved avionics and electronic systems. This was followed on 25 July 1990 with the first launch of an Atlas I, the last of the classic Atlas designs. The last of 11 Atlas I was launched on 25 April 1997, three of which failed to place their payloads in the correct orbit.

A total of 23 Atlas IIA/AS were launched from LC-36B between 7 December 1991, introducing the type to service from LC-36, and 19 May 2004. Introduced with a launch on 24 May 2000, the Atlas III version adopted a new first stage powered by a Russian RD-180, the first Atlas to dispense with the traditional 1½ stage configuration where the two booster motors are jettisoned leaving the central sustainer to continue firing until burnout.

The RD-180 more than doubled the first stage thrust of the Atlas rocket and greatly increased the payload capability. The IIIA version had a single RL-10 cryogenic rocket motor in the Centaur stage with the IIIB having the conventional twin RL-10 configuration. The last of six Atlas III was launched on 3 February 2005, all from LC-36B and all successful, an event also marking the last use of LC-36 pads.

BELOW LC-36B shortly after it was commissioned showing the classic Atlas launch pad configuration and layout for supporting Centaur upper stage configurations. *(NASA)*

BELOW RIGHT Atlas I lifts off from LC-36B, a succession of Atlas Centaur, Atlas IIA/AS types flown between 1965 and 2005, after which both LC-36A and LC-36B were retired. *(NASA)*

LC-41 (21 December 1965)
28.5834 °N x 80.5829°W

Along with LC-40 this complex was set up as part of a dual launch facility fed by the Integrated Transfer Launch (ITL) concept and details of this and the rationale behind the pad can be found in the description of that site.

The first launch from LC-41 of a Titan IIIC/Transtage went well until the terminal stage malfunctioned, placing the satellites it carried in the wrong orbit. The next launch on 16 June 1966 set a record for the eight satellites it successfully placed in geosynchronous orbits. These military communication satellites were part of multiple clusters which also formed the payloads for the third launch on 26 August and the fourth on 18 January 1967. Successive launches placed a variety of defence-related satellites in space, including more clusters of small communication satellites.

NASA utilised LC-41 for the first time on 10 December 1974 for the launch of Helios 1, an astronomy satellite developed jointly with West Germany but also as a proof test for the new Titan IIIE/Centaur, the first of which, launched on 11 February that year, had failed following a malfunction in the Centaur stage. Helios 1 was followed by two highly important NASA missions, Viking 1 on 20 August 1975 and Viking 2 a few weeks later on 9 September, both of which were highly successful in sending orbiter/landers to the planet Mars, achieving the first soft landing and surface-sampling activity.

Helios 2 followed on 15 January 1976 and then another two back-to-back flagship missions for NASA: Voyager 2 on 20 August and Voyager 1 on 5 September. Voyager 1 accelerated past Voyager 2 and flew past Jupiter in March 1979 and Saturn in November 1980 before heading for interstellar space; Voyager 2 flew by Jupiter in July 1979, Saturn in August 1981, Uranus in January 1986 and Neptune in August 1989 before it too exited the solar system.

These were the final launches of the Titan III series before modifications to the pad to prepare it for a new generation of more powerful Titan variants. Changes included refurbishing the Mobile Service Tower, the Umbilical Tower and various support facilities around the complex. There were also significant modifications to the Vertical Integration Building (VIB). The first launch of the powerful Titan IV series took place from LC-41 on 14 June 1989. A further evolution of the basic Titan IIIC, this launcher had more powerful boosters delivering a lift-off thrust of 3.2million lb (14,234kN) and the option of a Centaur third stage. A total of nine Titan IV-A were launched from LC-41, the last on 12 August 1998, with the sole Titan IV-B flown on 9 April 1999.

While Titan flights continued from LC-40,

FAR LEFT A Titan IVA/Centaur carrying the MERCURY electronic intelligence gathering satellite at LC-41. *(USAF)*

LEFT LC-41 supports the launch of a NASA Tracking and Data Relay Satellite from a modified facility supporting Atlas V. *(NASA)*

a new era was planned for LC-41 with reconstruction of the complex for the new Atlas V series in the Evolved Expendable Launch Vehicle (EELV) programme managed by the US Air Force. Demolition of the old facility began on 14 October 1999 when the Olshan Demolishing Company blew up the Umbilical and Mobile Service Towers. During the summer of 2000 a new Vertical Integration Facility (VIF) was erected 1,800ft (549m) to the south of the pad, a structure 292ft (89m) tall and equipped to stack the Atlas V on its Mobile Launch Platform (MLP). Consisting of a 5ft (1.5m) thick slab it rested on 65ft (19.8m) of pilings to support the great mass of the rocket and the VIF.

Nearby, a single 42,000 US gallon (158,970 litre) liquid hydrogen tank provided cryogenic fuel while two 45,000 US gallon (170,325 litre) liquid oxygen tanks provided cryogenic oxidiser for the Atlas V launch vehicle. Construction of the MLP was completed in 2001 as was the new Atlas Spaceflight Operations Center (ASOC) which included a mission operations room, a two-storey amphitheatre, a two-storey launch control centre and a range of supplementary support rooms. In all, 30,000ft² (2,787m²) of floor space. The ASOC could process up to six Atlas V launchers simultaneously.

The first Atlas V was launched from LC-41 on 21 August 2002, a civil communications satellite, the first of a series of commercial launches before NASA used LC-41 to send up the Mars Reconnaissance Orbiter spacecraft which successfully began orbiting Mars in March 2006 and at the time of writing (July 2018) is still operating. Yet another milestone event for NASA was the launch from this complex of the New Horizons spacecraft on 19 January 2006, the last robotic emissary to fly by a previously unexplored planet in the solar system, encountering the dwarf planet Pluto in July 2015, largest of the Kuiper Belt objects, targeted for a fly-by of another dwarf planet in January 2019.

After a series of launches for defence and civil commercial customers, NASA used LC-41 again for the launch of Juno on 5 August 2011 which entered an orbit of Jupiter in July 2016 to study this giant planet from a near polar orbit, thus observing for the first time phenomena at very high latitude. This flight was followed by the launch of Mars Science Laboratory and the Curiosity lander on 26 November 2011. The size of a small SUV, the Curiosity lander was lowered to the surface of Mars from a hovering sky crane on 6 August 2012.

Unexpectedly, commercial cargo carrier Orbital ATK (now a part of Grumman Innovation Systems) booked a flight of its Cygnus module on an Atlas V shortly after its Antares launcher exploded on the pad at Wallops Island, Virginia. Seeking a means of sustaining its contractual commitment to launch cargo to the ISS, and to honour manifested cargo already being prepared for the station, Orbital ATK booked slots to carry Cygnus until launches from Wallops Island could resume. The first Atlas V carrying a Cygnus (CRS-4) was launched from LC-41 on 6 December 2015 followed by CRS-6 on 23 March 2016 and CRS-7 on 18 April 2017.

Through this decade LC-41 has hosted a variety of satellite and space vehicle launches, mostly defence-related payloads, and will continue to do so. But in this new and rapidly evolving era of space commercialisation, existing facilities are being leased for non-government programmes and Boeing is developing its CST-100 Starliner, a ballistic capsule with which it hopes to deliver astronauts to the International Space Station from US soil for the first time since the last Shuttle launch on 8 July 2011.

Optimistically, Boeing has reserved a launch slot for the Starliner from LC-41 on 27 August 2018. If not achieved on that date there is no doubt that it is from this site that the Boeing crew-carrying spacecraft will, eventually, depart for the ISS.

LC-39A (9 November 1967)
28.6082°N x 80.6040°W

When originally laid out, LC-39 was to have consisted of five pads; the first three pads were to have been assigned suffix letters A, B and C in reverse order to that which they carry today running south from Playalinda Beach to Titusville. The existing LC-39A actually built is the pad originally designated LC-39C while the LC-39B actually built is the originally designated LC-39B. The original LC-39A to the north of it, as well as D and E still farther to the north were never built. But all that ended with significant pressure placed on the NASA budget in 1964 and the only two pads built at this complex were LC-39A and LC-39B identified as such from the south to the north.

The pads are virtually identical and this description will suffice for LC-39B too. Each takes the form of an eight-sided polygon and covers an area of about 0.25mile² (0.647km²) and the two pads are 1.65 miles (2.66km) apart. As measured by the odometer on vehicles moving along the Crawlerway, LC-39A is 3.44

miles (5.53km) from the VAB. Consideration of the azimuthal orientation of the ascending rocket dictated the angular alignment of the pad but it was the flame trench that coincidentally points precisely due north.

The distance across the pad is 3,000ft (914m) while the hardstand area is 390ft (118.8m) by 325ft (99m) and consists of several major elements, the pad itself including the hardstand, a Pad Terminal Connection Room (PTCR), the Environmental Control System (ECS), the High Pressure Gas Storage Facility (HPGSF), the flame trench and apron and the Emergency Egress System (EES). Each pad would require 4,627 tonnes of reinforced steel and 68,000yds³ (52,000m³) of reinforced concrete.

Each pad was built under a separate contract, work on LC-39A starting in November 1963 and essentially completed on 4 October 1965. The AS-500F facilities checkout vehicle, a Saturn V structural replica to verify the interface between attachments, connections and hold-down arms as well as the clearances of the swing arms on the Launch Umbilical Tower,

BELOW The general layout of LC-39A (foreground) and LC-39B, the only two pads where once at least five were considered in the very early days when NASA was thinking that up to 12 to 15 Saturn V flights a year might be feasible. *(NASA)*

was rolled out to LC-39A for the first time on 25 May 1966 – five years to the day that President Kennedy declared the Moon goal.

A major problem was compensating for the swampy conditions at the site and a surcharge of dredged fill from the Banana River some 80ft (24.4m) high was necessary to consolidate the bottom soil strata. It took the form of a flattened stepped platform to provide weight and compression for consolidating the undersoil. For each "pyramid" there was 500yds³ (382.3m³) of fill weighing a total of 680 tonnes. This settled the area by 4ft (1.22m) before the fill was removed for construction to begin. The floor on to which the flame trench and the pad hardstand was constructed was a concrete mat, 11ft (3.35m) thick, 150ft (45.72m) wide and 450ft (137.1m) long.

As built, with the floor of the flame trench level with the surface, the top of the pad on to which the Crawler Transporter would deliver the Saturn V on its Mobile Launch Platform was 48ft (12.8m) above the ground. The pad itself would consist of cellular structures, either side of the trench, to support the load. Each was 400ft (121.9m) long by 40ft (12.19m) wide and 42ft (14.6m) high. They consisted of individual cells with concrete diaphragms at 20ft (6.1m)

centres the walls of each cell about 3ft (0.91m) thick and the diaphragms 33in (93.8cm) thick.

The operational concept of the mobile launch system was to carry on the MLP all the interfaces necessary between the ground and the Saturn V through umbilicals and swing arms as described earlier in the Moonport section. Thus connected, the entire structure would roll out of the VAB to the launch complex where it would plug in to electrical, propellant, pneumatic and hydraulic systems carried permanently on nine service towers to interface with the Mobile Launch Platform. In addition there were six pedestals on to which the Mobile Launch Platform and its Saturn V would be lowered by the Crawler Transporter. During launch, two additional extendible arms temporarily fastened the bottom of the MLP to the pad to take the dynamic loads and the rebound which could reach 4,7562 tonnes.

The flame trench was 450ft (137.1m) long by 58ft (17.67m) wide and 42ft (12.8m) high, essentially the sides of the two parallel cellular structures. The floor and walls were covered with a refractory brick which could withstand temperatures of up to 1,670°C (3,038°F). The inverted V-shaped flame deflector would channel the flame fore and aft along the trench and was 41.5ft (12.65m) high, 48ft (14.63m) wide and 77.25ft (23.54m) long and weighed 2,866 tonnes. It was moved on rails in and out of the flame trench as required, with a second deflector available on a rail spur north of the pad.

Built of reinforced concrete, the PTCR consisted of two storeys situated on the west side of the flame trench beneath the sloping shoulder of the pad, covered with up to 20ft (6.1m) of earth fill. Inside, it housed electrical and electronic equipment which provided a link for power and data via transmissions lines from the Launch Control Center (LCC) and the Mobile Launch Platform. On the same side of the pad and similarly buried was the ECS room serving as a distribution point for conditioning and purge gases. Pressurisation of electrical and water systems was provided by nitrogen and helium in the HPGSF beneath the pad on the east side of the flame trench.

The Emergency Egress System consisted of a personnel escape system for pre-launch emergencies and consisted of a stainless chute,

BELOW Seen here during construction, the shape of the flame trench allows the free flow of hot gases in opposing directions vectored by a flame deflector, with the parallel sides to support the Crawler Transporter. (NASA)

about 200ft (61m) long in a superelevated curve which started at the interface with the Mobile Launch Platform and terminated 40ft (12.2m) below the pad in a rubber-lined isolation room. An adjacent blast-resistant room contained 20 contoured couches and safety harnesses and survival equipment for up to 24 hours. The dome-shaped blast room was 40ft (12.2m) in diameter with 2.5ft (0.76m) thick walls capable of withstanding a blast pressure of 500lb/in^2 (3,447kPa). A floating concrete floor supports the chairs on a suspension system with reduces the 75g force to 4g.

An industrial water supply system served both launch areas with a pumping station capable of furnishing water at 45,000 US gallons/min (170,325 litres/min). Two electrical distribution systems were provided, one for industrial power and the other for instrumentation, with emergency generators for added reliability. Support infrastructure included storage facilities for the propellant which would be used in the Saturn V and later in the Shuttle and other launch vehicles. Liquid oxygen was stored in a sphere with a 900,000 US gallon (3,406,500 litre) capacity located 1,450ft (442m) from the pad with a transfer line attached allowing a flow rate of 10,000 US gallons/min (37,850 litres/min).

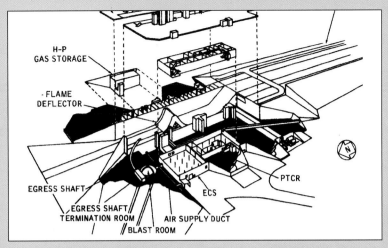

Liquid hydrogen was supplied from an 850,000 US gallon (3,217,250 litre) vessel also 1,450ft from the pad, delivered at a maximum rate of 2,000 US gallons/min (7,570 litres/min) through a 10in (25.4cm) pipe. Only the first stage of the Saturn V used a hydrocarbon fuel, RP-1, a course-grain kerosene, contained in three 86,000 US gallon (325,510 litre) capacity tanks 1,350ft (411.5m) from the centre of the pad and situated on the east side. This could be pumped through an 8in (20.3cm) pipe at a rate of 10,000 US gallons/min (37,850 litres/min). All propellant operations were controlled from the Launch Control Center.

ABOVE A partial cutaway of the pad with integral rooms and control equipment. *(NASA)*

BELOW The various pedestal and utilities supply conduits to mate the pad facilities with the Mobile Launcher. *(NASA)*

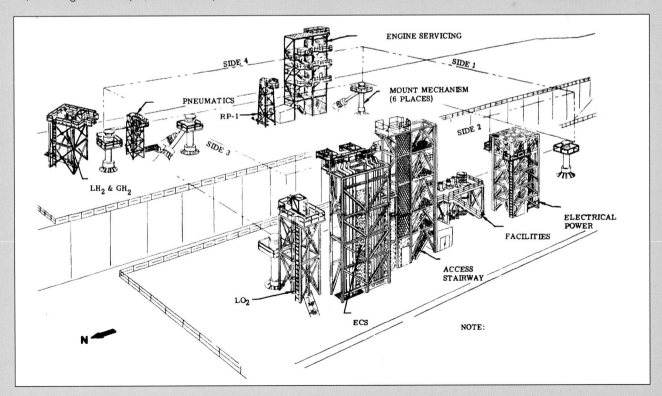

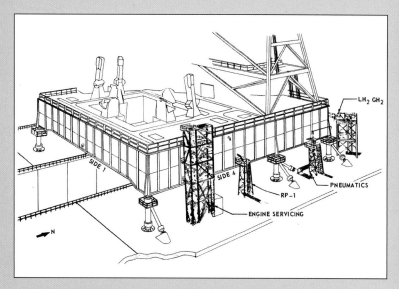

The Moonport facilities were one of the nation's biggest engineering triumphs, completed on time and with phenomenal success to remain today one of the most enduring of the space facilities constructed during the Space Race and still in service. Considered oversized when NASA made it to the Moon after only six of the 15 Saturn Vs thought necessary in 1962, LC-39 is now one of the strongest assets on hand and is the departure place for NASA as well as commercial rockets.

The first Saturn V to make it to LC-39A was not a rocket at all but a facilities checkout vehicle (AS-500F), which rolled out to the pad on 25 May 1966. The first launch of a Saturn V (AS-501) occurred on 9 November 1967, vindicating the all-up systems testing introduced with the management changes in 1963 and 1964, proving that high levels of quality assurance could circumvent the need to fly large numbers of launches gradually introducing upper stages into the flight sequence rather than launching from the outset, all in one stack. Designated Apollo 4, it threw an Apollo spacecraft to an apogee of 11,344 miles (18,256km), demonstrating a successful performance for the spacecraft and its heat shield.

AS-502 (Apollo 6) followed on 4 April 1968 but vibration problems caused an early shut-down of two of the five engines in the S-II second stage which resulted in the stage continuing to fire for an additional 58 seconds, requiring the S-IVB third stage to burn 29 seconds longer to make up the velocity

deficiency to achieve orbit. When it came to firing the third stage a second time to push the Apollo spacecraft on an elliptical path similar to Apollo 4, it failed to fire. Instead, the Apollo Service Module fired for 7min 22sec to reach an apogee of 13,832 miles ((22,259km) and the Command Module was returned safely to Earth, although there was insufficient propellant remaining to push it to a higher velocity in further test of the heat shield.

Despite the technical problems of its predecessor, AS-503 was launched on 21 December 1968 on one of NASA's most audacious missions. Carrying astronauts Borman, Lovell and Anders (on only the second manned Apollo mission) out of Earth orbit for the first time in human space flight, it placed the Apollo 8 spacecraft in lunar orbit on Christmas Eve. In one of the more moving episodes in human space flight, the crew each read a verse from the first chapter of Genesis as TV was broadcast showing the lunar surface to more than a billion people in 64 countries. They departed for Earth on Christmas Day and splashed down in the Pacific on 27 December.

Following the unmanned flight of the first Lunar Module in a Saturn IB on 22 January 1968, AS-504 carried a habitable LM and a manned Apollo spacecraft lifting the Apollo 9 crew of McDivitt, Scott and Schweickart into Earth orbit on 3 March 1969. The Saturn V performed as expected and the mission provided the first all-up test of both the Apollo and the LM. Extracted from its adapter on top of the S-IVB stage, the LM was flown on a sequence of separation and rendezvous manoeuvres, validating the engineering of both the LM and its propulsion systems. It gave Schweickart the first Apollo spacewalk when he demonstrated the ability of an astronaut to leave via the front porch and make his way to an open hatch on the Command Module in the unlikely event that neither spacecraft could dock with the other on returning from the lunar surface.

The fifth Saturn V launch took place from LC-39B (which see) but the next flight from LC-39A carried AS-506 into orbit on 16 July 1969 and the Apollo 11 crew of Armstrong, Collins and Aldrin to the first manned landing on the Moon four days later. The Moon landing goal had been achieved within eight years and two months of the historic goal set by President Kennedy and the crew successfully returned to Earth on 24 July.

To this date the pace had been remorseless, with Saturn V launches every two months or so and if Apollo 11 had failed to land there had been a back-up plan to launch AS-507 in September. That was relaxed and Apollo 12 flew on 14 November 1969 carrying Conrad, Gordon and Bean to the second lunar landing, a pin-point touch-down close to the previously launched Surveyor III spacecraft and a double EVA before returning to Earth. To reduce costs and to allow results from one mission to feed into another, the third attempted Moon landing began with the launch of AS-508 on 11 April 1970 but one of the S-II stage engines shut down early, without compromising the Apollo 13 mission. However, one of two oxygen tanks exploded imperilling the lives of the crew, who used the LM for a substantial part of the circumlunar return to Earth and safe splashdown.

The next launch was delayed until 31 January 1971, when AS-509 carried the Apollo 14 crew of Shepard, Roosa and Mitchell to NASA's third Moon landing, followed by AS-510 with Apollo 15 (Scott, Worden and Irwin) on 26 July 1971, AS-511 with Apollo 16 (Young, Mattingly and Duke) on 16 April 1972 and AS-512 with Apollo 17 (Cernan, Evans and Schmitt) on 7 December 1972. These last three missions carried far greater capacity for remaining on the surface of the Moon,

supporting three full Moonwalks and scientific surveys supported by a Lunar Roving Vehicle, each of which remained on the lunar surface.

The final launch from LC-39A in the Apollo era was that of AS-513, a two-stage Saturn V launched on 14 May 1973 carrying the pre-fitted Skylab space station, a converted S-IVB stage. During ascent air pressure built up in a duct and ripped away the meteoroid shield on Skylab, unlatching one of two solar array booms previously held tight against the hull of the station. Also, the adapter between the S-IC and S-II stages failed to jettison after first stage separation, causing heat to build up with the five S-II engines firing and almost causing a complete failure before finally reaching orbit. When the retro-rockets fired on the adapter connecting the S-II to Skylab the forward-firing blast tore away the freed solar array wing.

After the last Saturn V had been launched work began to adapt it for the Shuttle programme and this would require substantial

BELOW The Launch Umbilical Tower and swing arms provided connection to each stage and each spacecraft. *(NASA)*

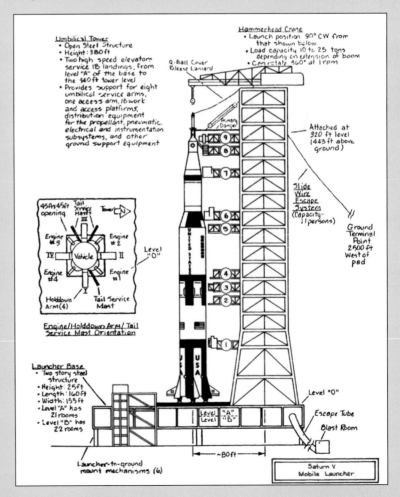

RIGHT The functional purpose of LC-39, the flight of a Saturn V, in this case lift-off for the ill-fated Apollo 13 mission. *(NASA)*

FAR RIGHT The flame deflector in place between the concrete cells in the launch pad. *(NASA)*

BELOW The configuration of LCX-39A/B pads changed completely with the Shuttle, the LUT and the Mobile Service Structure being removed and a permanent servicing and access gantry built on to the side of the pad. The rotating section containing the Payload Changeout Room can be seen at the left, the entire structure hinged to move round and encapsulate the Shuttle along a curved rail track seen just below the Crawler Transporter. *(NASA)*

changes to the pad. More than three years in to development of the Shuttle, NASA was confident it would be launching upward of 45 flights a year with this reusable transportation system and the ability to fly from each pad once a fortnight stressed the capacity of the infrastructure built for Saturn V and the Apollo programme. Of course, such a high flight rate was never achieved and that avoided an almost insoluble challenge including all the Moonport facilities as well.

A period of almost eight years would elapse before thunder would once again roll across the Cape from LC-39A, the first Shuttle flight occurring on 12 April 1981. In the interim, a considerable amount of rework was carried out to configure the pad for the winged spaceplane. The Shuttle would carry two Solid Rocket Boosters (SRBs) as well as three cryogenic main engines at the base of the Orbiter and the overall height of the vehicle was only half that of Saturn V so the Launch Umbilical Tower was substantially modified, that being described in the Moonport USA section.

Both pads would be reconfigured to provide for servicing at the pad, where the payload would be placed inside the Orbiter. This was very different to the Apollo era where the Apollo spacecraft was stacked inside the VAB and the Launch Umbilical Tower was situated alongside the Saturn rocket on the Mobile Launch Platform before it was wheeled out. The Mobile Service Structure constructed for the Apollo era was no longer required for the Shuttle, replaced by a Fixed Service Structure (FSS) at the pad itself to which was attached a Rotating Service Structure (RSS), hinged to move round once the Shuttle was in place and the MLP on its legs at the pad to encapsulate the Orbiter.

The FSS was constructed on the west side of the pad, essentially an open framework structure 40ft (12.19m) square which supported the hinge about which the RSS pivoted. The FSS had a total height of 247ft (75.3m), with a hammerhead crane taking the height a further 18ft (5.5m) and a lightning mast carrying the total height to 347ft (105.76m) above the pad. The FSS had 11 levels, at 20ft (6.1m) intervals from the first at 27ft (8.2m) above the pad. Just three service arms were installed, a gaseous oxygen vent arm, a hydrogen vent line and access arm and an Orbiter access arm.

The Orbiter access arm swung out 70° from the FSS to provide access to the Orbiter side hatch for getting in or out of the Orbiter, with a "white" room capable of supporting six people. The arm was 147ft (44.8m) above the pad and was 65ft (19.8m) long, 5ft (1.52m) wide and 8ft (2.4m) high. Next up was the gaseous hydrogen vent and intertank access and umbilical arm, 48ft (14.6m) long and weighing 15,800lb (7,167kg) which rotated 210° to its extended position. The gaseous oxygen vent arm was 65ft (19.8m) long from the tower hinges to the vent hood hinge. The diameter of the vent hood (affectionately known as the "beanie cap") was 13ft (4m) in diameter. The 80ft (24.38m) fibreglass lightning mast was grounded by a cable anchored 1,100ft (335m) south of the pad to 1,100ft (335m) north, angles up and over the mast to act as an electrical insulator holding the cable away from the FSS.

An emergency escape system was built in to the FSS so that crewmembers could rapidly escape an impending explosion or other disaster by descending from the Orbiter access arm along seven slidewires extending from the level of the access arm to the ground on the west side of the pad. A single three-person basket is attached to each slidewire and restrained by a brake until released, whereupon the occupants slide down the 1,200ft (365.7m) wire to arrestor wires at the landing zone and access to underground bunkers.

The Rotating Service Structure was supported by a rotating bridge that pivoted about a vertical axis on the west side of the pad flame trench and had a pivoting arc of 120° on a radius of 160ft (48.77m). Support for the outer end of the RSS is by twin eight-wheel trucks that move along twin rails installed flush with the pad surface, crossing the flame trench on a curved bridge.

The RSS was 102ft (31m) long by 50ft (15.24m) wide by 130ft (39.6m) high. The bottom of the main structure was 59ft (18m) above the surface of the pad, the top at a height of 189ft (57.6m). It had Orbiter access arms at five levels to afford access to the Orbiter payload bay while also affording access to other levels of the vehicle. A fundamental function of the RSS was to provide a receiving station for environmentally protected payloads which had been kept in near-sterile conditions where they had been built, constructed or assembled, and delivered to the interior of the Orbiter payload bay in an equally contamination-free condition. This introduced two elements

BELOW The general arrangement and measurement of the fixed and rotating service structure. *(NASA)*

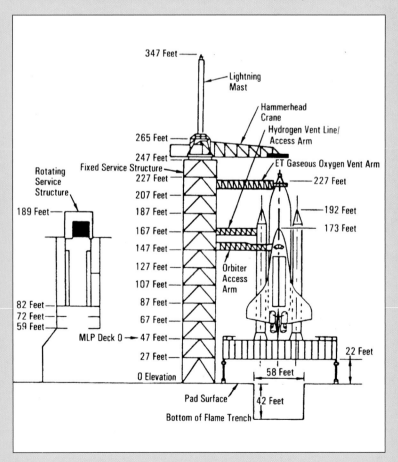

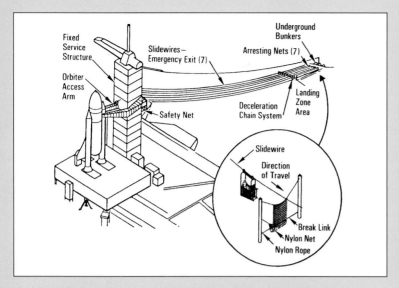

to make that possible, a Payload Canister (PC) to bring the payloads to the pad in a protected condition and a Payload Changeout Room (PCR) on the RSS itself to allow the payloads to be transferred to the Orbiter.

The Payload Canister (PC) had an interior which was the size of the Orbiter payload bay at 65ft (19.8m) by 15ft (4.57m), and a total external dimension of 65ft (19.8m) long by 18ft (5.48m) wide and 18.6ft (5.6m) high. It had a carrying capacity of 65,000lb (29,483kg) – the maximum

design capacity of the Shuttle but one never actually achieved. The PC was moved around on a flatbed truck 65ft (19.8m) long and 23ft (7m) wide with 48 wheels and capable of being raised to 7ft (2.1m) or lowered to 5.25ft (1.6m). The transporter had an unloaded weight of 104 tonnes and was capable of being steered from either one of two diametrically opposed cabs from a crawl of 0.25in/sec (0.6m/sec) to a maximum speed of 10mph (16kph).

As an integral part of the RSS, the Payload Changeout Room was an environmentally controlled space matching the dimensions of the Orbiter payload bay so that it could be hermetically sealed and treated as a "white" room to minimise contamination. With the RSS in the open position, the Payload Canister would be winched up to the interior of the PCR. Once the RSS closed with the Canister encapsulated, the Orbiter's payload doors could open and payloads transferred across to the interior of that bay in a situation which would prevent contamination from the coastal environment.

Some massive payloads such as structures for the assembly of the International Space Station or pressurised modules either retained in the Shuttle for the duration of the mission, or removed and attached to the ISS, would be processed horizontally, for which the Orbiter Processing Facility (OPF) was built, described in the section Moonport USA (see page 113). Payloads installed here were fitted inside the Orbiter before it was rolled around to the VAB for vertical assembly.

An addition to the LC-39 facilities was made necessary after the first launch of the Shuttle when high sound levels from the Solid Rocket Boosters battered the Orbiter, loosening some of the thermal tiles. The standard sound suppression system designed in from the outset took the form of 16 nozzles on top of the flame deflectors connected to a 300,000 US gallon (1.135million litres) tank 290ft (88.4m) north-east of the pad from which water began to spray just before ignition of the three main engines. Under the modified sound suppression system, when the solid boosters ignited a torrent of water flowed from six large nozzles known as "rainbirds". Water was also sprayed into the exhaust holes for the solid boosters to provide

BELOW Dimensions of the flame deflector which during Apollo was surfaced with a volcanic ash aggregate, turned to glass in the heat of the F-1 engines of a Saturn V. *(NASA)*

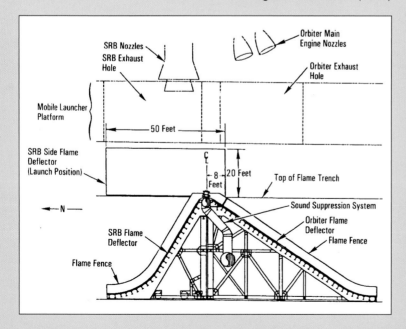

RIGHT The sound suppression water system designed to deaden acoustic waves which buffet the trailing edge of the wings, the flight control surfaces and the tail unit of the Orbiter, dislocating thermal protection tiles and causing damage. (NASA)

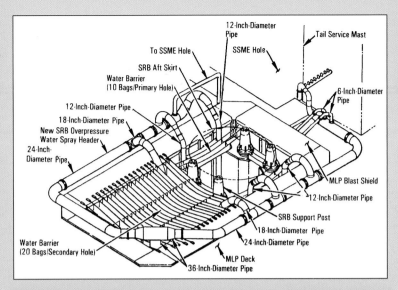

overpressure protection with peak flow reaching 900,000 US gallons/min (3.4million litres/min) about nine seconds after lift-off.

LC-39A supported 83 Shuttle launches, including the first, STS-1, on 12 April 1961 and the last on 9 July 2011 when STS-135 took *Atlantis* on its last mission to the International Space Station. Nearly eight years separated the launch of the last Saturn V and the first flight of the Shuttle and almost six years separated the flight of the last Shuttle and the next use of LC-39A. When Shuttle flights ended it had been expected that a new rocket, the Ares V, a derivative of legacy Shuttle hardware, would begin flying in support of a new generation of Moon explorers defined by the Constellation programme.

When Barack Obama cancelled Constellation in 2010 it delayed activity at the Cape and created a vacuum where 6,000 jobs were lost and only a few contractor personnel went elsewhere. Support for commercial launch providers had been growing, however, and there was increased interest in marketing the facilities for lease to these companies. Over the next several years, companies such as SpaceX bargained a lease on LC-39A, from where today Elon Musk launches his Falcon 9 and where the Falcon 9 Heavy was launched, sending a Tesla car to solar orbit.

The first launch from LC-39A after the final Shuttle mission was the Dragon CRS-10

CENTRE A test of the deluge when the sound suppression water system was activated. (NASA)

RIGHT On 14 April 2014, SpaceX CEO Gwynne Shotwell signed a contract for the use of LC-39A for Falcon 9 and Falcon Heavy launchers, the first commercial flight of that pad coming on 10 February 2017 with a Falcon 9 carrying cargo to the International Space Station. (NASA)

mission on a Falcon 9 v1.2 rocket, the first stage returning to a controlled descent and a soft landing on LZ-1 (see LC-13, page 149). Further commercial launches followed, with the Falcon Heavy on 6 February 2018, the two side-boosters returning to a synchronised soft-landing at LZ-1 and LZ-2. Crewed flights using the Dragon 2 spacecraft are likely to begin from LC-39A in 2019.

LC-25C (16 August 1968) 28.4311°N x 80.5764°W

One of four launch pads to support America's SLBM programme, LC-25C was one of two in this complex built to support test launches for the Poseidon missile, larger than Polaris and requiring a dedicated launch site of its own. The blockhouse supporting all four LC-25 pads was extensively modernised for the new missiles.

LC-25C supported 16 Poseidon C3 launches, the last on 30 June 1970. Significant work was conducted to prepare the pad for test flights with the Trident C4 missile, of which 18 were fired between 18 January 1977 and 23 January 1979. The pad was deactivated and dismantled in 1979.

LC-39B (18 May 1969) 28.6272°N x 80.6208°W

Construction of this pad began in December 1963 and was completed on 30 November 1966. Essentially identical to LC-39A, it differs most noticeably in that the top of the pad is 55ft (16.7m) above sea level. Specific details can be found in the entry for LC-39A (page 177). Construction of LC-39B began in December 1963 and was completed on 30 November 1966 but it was not until the Apollo 10 flight of 18 May 1969 that it was first used, the only time it was for the flight of a Saturn V for which it had been built.

The next launch occurred on 25 May 1973 for SL-2, the second Skylab launch and the first to carry a crew to the Skylab space station, which had been launched from LC-39A 11 days before. SL-2 involved the use of the Saturn IB, which had previously been launched from LC-34 or LC-37. For cost reasons both these facilities had been closed but the need to support three Skylab crew flights from LC-39 required modifications that involved the use of what was euphemistically called the "milkstool"!

This was a significant modification to Mobile Launcher 1 to adapt it for the Saturn IB/Apollo configuration, which was virtually identical to the Saturn V from the S-IVB and payload up. It was decided to raise the upper elements to the same relative height they would occupy were they on that bigger rocket by erecting a steel pedestal (the "milkstool") to allow the upper stage and payload to be raised to the height that would occupy were they attached to a Saturn V and that they were compatible with the swing arms on the Launch Umbilical Tower. The pedestal stood 127ft (39m) tall and tapered from 48ft (14.6m) square at the base on ML-1 to 21.9ft (6.6m) at the top where it supported the first stage. Installation required removal of the Saturn V hold-down arms, service arms 1, 2 and 3 and four tail service masts.

In this configuration, ML-1 also supported the launch of SL-3 on 28 July 1973 and SL-4 on 16 November 1973 and was utilised again for the joint Apollo Soyuz Test Project, which saw the last Apollo hardware used operationally for a docking with a Soviet Soyuz spacecraft following launch on 15 July 1975. After the launch of ASTP work began to configure LC-39B for the Space Shuttle programme. Modifications almost identical to those described for LC-39A prepared for the first Shuttle launch (STS-51L) which occurred on 28 January 1986, the ill-fated loss of *Challenger*.

It was from LC-39B that Shuttle flights resumed with the launch of STS-25 (*Discovery*) on 29 September 1988. In total, 53 Shuttle flights began from LC-39B, the last being the night launch of *Discovery* for STS-116 on 10 December 2006. In May 2009 LC-39B supported the Shuttle *Endeavour* in the event that it was needed for a rescue mission to the Hubble Space Telescope, the STS-125 flight launched from LC-39A and the last mission for which a Shuttle was not docked to the ISS where the crew could reside for some time to await a rescue if needed.

Immediately after the return of STS-125, modifications began for flights from LC-39B in support of the Constellation programme designed to send astronauts back to the Moon. Two launch vehicles were being developed for that programme, hardware evolved from the Shuttle programme involving the powerful Solid Rocket Booster. Known as Ares I-X, it consisted of a single SRB with a simulated upper stage in a test of the configuration which would be used to place an Orion spacecraft in orbit. Launched on 28 October 2009, it was the last flight from this pad.

After this flight NASA returned LC-39B to the pre-Shuttle configuration with all the Fixed Service Structure removed in a bid to make it adaptable for any operator likely to use the facility in the future, for which opportunities opened in 2014. While the historic LC-39A site hosts SpaceX and Falcon 9 flights, LC-39B will support flights of NASA's giant Space Launch System (SLS) beginning with the first unmanned launch probably in 2020. The first manned SLS/Orion flight is likely to occur no earlier than 2022 by which time it is expected that commercial

WHAT'S IN A NAME?

Over these busy years the launch sites at Cape Canaveral and the Kennedy Space Center have gone through further name changes. Initially, on 11 May 1949 the site was signed over by President Truman as a launch range for guided missiles and a month later the Banana River Naval Air Station located 15 miles (24km) to the south of the Cape (transferred to the Air Force on 1 September 1948) was named the Joint Long Range Proving Ground.

To recognise the exclusive Air Force management of the site, on 16 May 1950 the Joint Long Range Proving Ground (LRPG) dropped the "Joint" followed a day later by the headquarters simply being known as the Long Range Proving Ground Base (LRPGB). In honour of Maj Gen Mason M. Patrick the LRPG was renamed Patrick Air Force Base on 1 August 1950.

On 30 June 1951 the Joint Long Range Proving Ground Division became the Air Force Missile Test Center (AFMTC) and the LRPGB became the Florida Missile Test Range, only to be renamed the Eastern Test Range in 1958 to allow for the establishment of the Western Test Range out of Vandenberg Air Force Base, California. Because of the importance placed on the missile work at Redstone Arsenal, Huntsville, Alabama, on 1 December 1951 the Experimental Missiles Firing Branch was established under the Army's Technical and Engineering Division.

Formed to supervise the building of Redstone launch facilities at Cape Canaveral Missile Test Annex, it secured facilities and support from the AFMTC. Following approval on 8 November 1955 for development of the Army's Jupiter IRBM, the Army Ballistic Missile Agency (ABMA) was formed at Redstone Arsenal and the Missile Firing Laboratory (MFL) became part of the ABMA's Development Operations Division which was effective from 24 December 1956. When this Division, under Wernher von Braun, had been transferred to NASA the MFL was renamed the Launch Operations Directorate (LOD) and this coexisted with the Air Force Missile Test Center (AFMTC).

When President Kennedy announced the Moon goal on 25 May 1961 and by implication the need for bespoke launch sites for heavy lift Saturn launch vehicles unlike anything built by the Air Force or the Army, NASA and the Department of Defense signed an agreement on 24 August for the acquisition of 80,000 acres (32,373ha). Located north and west of the Cape Canaveral Missile Test Annex NASA would build its assembly buildings, launch control facilities, launch pads and infrastructure here, the LOD becoming an independent NASA field installation effective 1 July 1962. At the Cape this would become the Launch Operations Center (LOC).

A week after President Kennedy was assassinated, on 29 November 1963 President Lyndon Johnson, who had done so much to create NASA and then persuaded Kennedy to commit to the Moon goal, renamed Cape Canaveral as Cape Kennedy and the Launch Operations Center as the John F. Kennedy Space Center (KSC). Later, citizens would lobby for the name Cape Canaveral to be revoked and it was so, leaving KSC at NASA's launch facility.

crewed flights to the International Space Station will have been flying from LC-39A.

LC-25D (17 September 1969) 28.4295°N x 80.5782°W

One of four pads built to support the SLBM programme for the US Navy, LC-25D was one of two supporting the more powerful Poseidon missiles. It supported only one launch, that of a Poseidon C3 on 17 September 1969, which reached a height of 311 miles (500km). Like LC-25C, the pad was deactivated and dismantled in 1979.

In 2012 the site for LC-25 and LC-29 was used for a new US Navy test facility known as the Strategic Weapons Systems Ashore, a single location for testing all the fire control, launch systems and navigation equipment in an integrated location, rather than having all the different contractors conduct those activities at their own facilities across the USA. It was completed in 2015.

LC-46 (15 January 1987) 28.4585°N x 80.5284°W

As built from February 1984, this complex was designed to support flight tests with the Trident II SLBM and it was accepted by the US Navy in November 1986. The Launch Control Center was situated in the Fleet Ballistic Missile Test Operations Facility, Building 62615 and this controlled the first launch on 15 January 1987. A total of 19 missiles had been launched by 26 January 1989, after which the Trident II moved to sea launches from March that year and LC-46 was placed on standby.

As no new Navy requirement emerged, the Spaceport Florida Authority won a US Air Force grant to adapt it for small commercial launches, several proposals for which were emerging in the early 1990s. New support infrastructure was built and the first launch occurred on 7 January 1998 when an Athena II launched NASA's Lunar Prospector. Two flights of this rocket had been launched from SLC-6 at Vandenberg Air Force Base, the first a failure, but this launch sent the spacecraft all the way to the Moon where it conducted a scientific survey from low polar orbit.

On 27 January 1999 an Athena I launched the Taiwanese satellite Rocsat-1 into orbit. On 15 July 2015 an announcement was made about refurbishment and modifications to the facility for launching the Minotaur IV small satellite launcher and on 26 August 2017 a Minotaur IV launched the ORS 5 defence technology satellite, the last launch to date.

RIGHT LC-46 was activated to test the Trident II ICBM, underpinning the sea-based nuclear deterrent for the United States and the United Kingdom. Note the unusual design of the blast defector tunnel. *(USN)*

FAR RIGHT Athena II was used by NASA to launch Lunar Prospector from a much modified LC-46 on 7 January 1998. *(NASA)*

Arguably the most famous and certainly most quoted NASA facility anywhere, the Johnson Space Center, named as such on 19 February 1973 in honour of President Johnson, started out as the Manned Spacecraft Center which, since 1965 during the two-man Gemini programme, has been responsible for operating all of the agency's human space flight activities. But its origin began long before then.

In fact, as early as 1960 NASA had been concerned about the lack of a single dedicated facility for supporting its manned operations. In that year only the one-man Mercury programme had been funded and that was a project inherited from the Air Force when NASA replaced the NACA in October 1958. Mercury had a lot of NACA input and the project found a natural home when the new civilian space agency took it over.

NASA was divided about the amount of emphasis that should be placed on human space flight, versus the burgeoning list of potential space science projects, not to mention planetary exploration with robots that fought for funding. While President Eisenhower was reluctant to pitch into full support for human space missions, preferring to wait and see what would come of this new endeavour, there were several senior managers at NASA that urged on the next step.

NASA had already determined that a sequential and expanding set of capabilities could be undertaken by a single space project which it named Apollo, rejected as a funded start by a conservative President but argued for by many who had been working as NACA employees to advance the day when humans left Earth. Nevertheless, there was a lot of conceptual work that had already prepared for the next steps beyond Mercury and that would entail a three-man vehicle capable of supporting Earth-orbit science, circumlunar and lunar orbiting flights as a prelude to a manned landing on the surface of the Moon.

At the time, and for the immediate future, NASA's Space Task group (set up by Administrator Glennan in November 1958), a component of the Langley Research Center, was the field element responsible for carrying out management and direction of operations leading up to the first US manned space flight when Alan Shepard made a ballistic flight lasting little more than 15 minutes on 5 May 1961. That flight gave the green light for President Kennedy to announce, exactly 20 days later, that America was heading for the Moon.

To give the manned flight programme a proper home, Administrator Glennan had already considered converting the Ames Research Center into NASA's manned flight facility and when Webb succeeded him in 1961 he asked for money to build a new centre but that was rejected by the Budget Bureau. Shortly thereafter, approval was granted and work began to find a suitable location, with the decision announced on 19 September that the facility would be built near Houston, Texas. There was speculation at the time that Texas had been chosen through the collusion of Vice President Lyndon Johnson, a Texan, and Houston Congressman Albert Thomas, chairman of the House subcommittee handling NASA's appropriations.

Personnel began moving out of offices in Langley from October 1961 and construction of

BELOW Home of NASA's human space flight programmes and mission management, the Johnson Space Center has the distinction of having given the agency some of the most outstanding successes in its history. *(NASA)*

the new facility began in April 1962. Officially, the
Manned Spacecraft Center opened for business
in September 1963 but the first operational
use of the facility to support a space flight was
the Gemini IV mission launch on 3 June 1965.
The responsibilities of the centre included
design, development and testing of spacecraft
and their associated systems for manned
flight, the selection and training of astronauts,
planning and conducting manned missions,
extensive participation in medical, engineering

and scientific experiments to assist with human
space flight and understanding the space
environment as it pertains to those activities.

To facilitate these responsibilities, the
organisation was divided into several
directorates in charge of specific functions, such
as spacecraft development, astronaut training
or space flight planning. A built-in flexibility
allowed frequent realignment to keep pace
with changing directions of manned flight, with
several of the directorates reorganised, merged
or split into separate groups, new directorates
being created as needed. The centre was also
responsible for directing local, agency-wide
or contractor support to fulfil its mission with
directorates and programme offices responsible
to the centre director reporting in turn to NASA
headquarters.

The 1,620 acre (655ha) site is about 25 miles
(40km) south-east of Houston alongside Clear
Lake on land donated by Rice University. It
consists of around a hundred buildings from
the nine-storey project management building
to tiny traffic booths at each entrance! A lot
of buildings contain office space but others
are dedicated to specific tasks and activities.
The development of this dedicated centre for
human space flight was approved during the
Eisenhower administration and before the big
decision made by President Kennedy to send
men to the Moon, despite the fact that initially
there was no money to build a new facility. But
the Moon goal made it imperative to set up this
new control centre.

Under the limited requirements of the
Mercury programme, NASA could manage
the flight from the Mercury Control Center at
Cape Canaveral but with the expansion of the
human space flight programme that became
impossible. It was logical that mission control
operations should move to the new Manned
Spacecraft Center but it was a decision made
as a result of the decision to create the Gemini
and Apollo programmes. The MSC would have
two Mission Operations Control Room (MOCR,
pronounced "moker") facilities – one to conduct
flight operations, the other to prepare for the
next mission. It was first tested during the
Gemini II flight of January 1965.

Probably the most famous building at the
Manned Space Center is the originally named

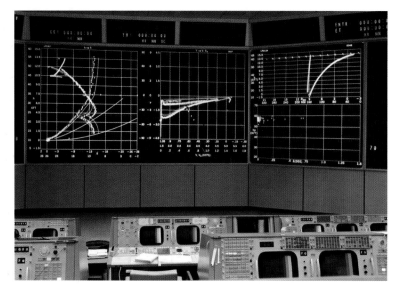

Integrated Mission Control Center (IMCC), more commonly MCC but officially on site as Building 30 which is split into 30-A, 30-M and 30-S. Building 30-A is responsible for the administration of the MCC and has historically been the place where flight and mission design takes place and where ground control operations are planned and coordinated. It is also the location of what was, in Apollo days, the Mission Planning and Analysis Division (MPAD).

The first flight controlled from the new Manned Spacecraft Center was Gemini IV launched on 3 June 1965, at which time the facility also had technical monitoring and development for both Gemini and Apollo programmes, the latter involving two separate spacecraft with very different engineering criteria. MSC was responsible for mission planning as well and for the training of astronauts. It was the home of the astronaut corps from which group were selected the men, later women too, who would fly specific missions. A range of new and untried operational procedures were managed by MSC including spacewalking, rendezvous and docking, the simultaneous communications, data handling and tracking of two manned vehicles in orbit, development of lunar mission activity, integration of science tasks, familiarisation and training with new generations of equipment and the development of future mission strategies.

As two-man operations involving a prime and a back-up crew during the Gemini years were replaced by three-man missions of Apollo with the addition of support crews, the number of astronauts assigned to a specific flight increased from four to nine and with up to four or five flights pre-assigned at the same time it meant up to 45 astronauts could be rotating through various missions, simultaneously vying for briefing schedules, training facilities and simulators. As far as possible tasks were polarised around a set of standard procedures but that did little to reduce the strain on personnel and timelines.

During Apollo, MSC was involved with applications studies for using the hardware to establish a permanent presence on the Moon or to use the existing hardware for more advanced missions and there was considerable

ABOVE While declaring "Task accomplished – July 1969" the scene in Mission Control at the successful conclusion of the Apollo 11 flight was but a prelude. With only 14 manned flights under their belt, the work would go on for decades to come managing missions on a sustained basis. *(NASA)*

development under way involving roving vehicles for carrying astronauts around on the lunar surface and various habitats for use as a base camp. That work disappeared as all hope of retaining the semblance of a sustained Apollo programme evaporated, replaced by studies into reusable shuttlecraft. Working in close cooperation with Langley and the Flight Research Center, MSC produced candidate designs for a low-cost, reusable space transportation system which would eventually become the Space Shuttle.

BELOW The Flight Control Room reconfigured for a very different function, seen here during the STS-128 mission in August 2009 just two years prior to Shuttle retirement. *(NASA)*

ABOVE **Above all the technology, it is the men and women of the Johnson Space Center who manage events and control the missions. Christopher Kraft was a leading architect of the Mission Control Center concept and the way it was set up for success, here directing operations from Cape Canaveral before moving to Houston.** *(NASA)*

LEFT **Updated and reconfigured again for sustained, continuous operation, the Flight Control Room has been supporting round-the-clock activity in space at the International Space Station since the late 1990s.** *(NASA)*

MSC was heavily committed between 1970 and 1975 with maintaining a flow of missions utilising Apollo hardware and developing the Shuttle, which saw many changes both to its size and configuration in that period. After the Apollo Soyuz Test Project, the joint flight with the Russians, undertaken in 1975, MSC was exclusively committed to the Shuttle, which had been formally approved in 1972, and entered a new era in which sustained operations with a manned vehicle would be the order of business for 30 years from the first Shuttle flight on 12 April 1981.

Astronauts were at the heart of the Manned Spacecraft Center with a changing role as the human space flight programme evolved. The first astronaut selection after MSC began formal mission management operation in 1965 were the fourth group selected by NASA since the first Mercury 7 in April 1959. These were also the first of NASA's science astronauts, a category that brought contention within the centre as many engineers, flight personnel and some managers feared that the missions were inappropriate for all but seasoned test pilots. Six science astronauts were chosen and three flew on the Skylab space station after one, Harrison H. Schmitt, had walked on the Moon during Apollo 17.

A further 19 astronauts were selected in April 1966, of whom three would walk on the Moon (Duke, Mitchell and Irwin) with many flying on Skylab or the Shuttle. The second group of scientist astronauts were selected in August 1967, 11 men who never expected to fly in space; with the demise of the Apollo Applications Program, no plans for follow-on manned flight programmes and an existing

LEFT **Support teams are a lifeline between events in space and flight management decisions in Mission Control, this team posing for a group shot during Apollo 7 in October 1968.** *(NASA)*

surfeit of astronauts on the team, they had little hope of a mission. In fact, with the Shuttle programme announced formally less than five years later, seven would make it into space. And as a portent of further decline, in August 1969 NASA inherited seven astronauts from the cancelled US Air Force Manned Orbiting Laboratory programme, all of whom flew on the Shuttle with one – Richard Truly – serving as NASA Administrator, the first astronaut to make it to NASA's top job.

Not for another nine years would NASA recruit any more astronauts, before 35 were selected in August 1969. Confident that it would get the Shuttle, much talked about in this year of the first Moon landings, the agency heralded a new era, one in which it chose to include six women. From this group would come the first American woman to fly in space (Sally Ride), the first African-American (Guion Bluford), and the first US woman astronaut to walk in space (Kathryn Sullivan). All 35 made it into space riding the Shuttle. Additional groups were selected at relatively frequent intervals, the latest being the 17th group selected by NASA in 2017 of whom six, exactly half, were women.

The astronaut corps has provided the space programme with some of its most spectacular achievements, although still only 339 Americans had flown in space as of January 2018, of which only 24 (of any humans anywhere) have travelled farther than low Earth orbit. Of those only 12 have walked on the Moon. To put this in perspective, only 24 people (of any nationality) have been farther from Earth than the 852 miles (1,372km) achieved by Gemini XI astronauts Conrad and Bean in 1966, less than the direct distance between London and Rome. This will change when NASA moves to deep-space operations in the early 2020s and once again astronauts travel to the gravitational environment of another world in space.

Known since 14 April 2011 as the Christopher C. Kraft Jr Mission Control Center in recognition of the outstanding work carried out by its first Flight Director, 30-M (for Main building) is the historic core of the Mission Control Center, also the historic heart of mission operations. Formerly known as Flight Control Room-1 (FCR-1) and now the ISS Control Room after an intermediate period as

the Shuttle FCR, it was originally the Mission Operations Control Room (MOCR-1) during the early years of Gemini and Apollo. It is the home of the Apollo Control Room and associated rooms including facilities for contractor personnel and a flight controller training room.

Originally known as the Space Shuttle Control Room, 30-S (designated for Space Station Freedom as the ISS was originally known) was renamed the White FCR before transitioning to the control room for the International Space Station. There is now a Blue FCR for the Orion spacecraft but the facility has seen several shared and exchanged responsibilities during the time when the Shuttle and the ISS were operationally concurrent and requiring separate control room support. It is now known as the Christopher C. Kraft Mission Control Center in honour of the man who made a strong reputation as the first flight director and then, from 1972 Director of the then Manned Spacecraft Center.

ABOVE Training is the key element for achieving success and it began early in the space programme with extensive preparations for the exploration of the Moon as conducted here by astronauts Shepard (right) and Mitchell prior to their Apollo 14 flight. *(NASA)*

In support of spacecraft development and engineering, Building 32 houses the Space Environment Simulation Laboratory (SESL), two vacuum chambers, one 120ft (36.5m) high and 65ft (19.8m) in diameter, the second 43ft (13.1m) in height by 35ft (10.7m) in diameter. Spacecraft or elements of a vehicle can be subject to vacuum conditions and temperatures between -250°F (-157°C) to +250°F (121°C). Contrastingly, Building 14 houses an Anechoic Chamber Test Facility where foam-covered walls, floor and ceiling soak up stray signals during spacecraft communication tests.

Building 9 has provided simulation facilities for full-scale mock-ups of hardware, from deployment of scientific instruments on the lunar surface during Apollo to Shuttle Orbiter crew familiarisation to ISS module familiarisation and training. Known as the Space Vehicle Mockup Facility (SVMF), it is a large open space with a 656ft (200m) long chamber that is used primarily to contain ISS modules for familiarisation training and will play an important role in the future as NASA and the ISS international partners prepare for supporting the Lunar Orbital Platform-Gateway, a collection of pressurised and unpressurised modules designed to provide sustained habitation in lunar orbit.

The need for simulating the space environment extended to humans and in early 1960 NASA recognised that the degree of neutral buoyancy effected by water was a good approximation of the weightless sensation in space. At first, NASA made use of a public pool near the Langley Research Center but when visitors disturbed the work it moved trials to a swimming pool at McDonogh School in Maryland. It was there that Malcolm Scott Carpenter became the first astronaut to carry out suited trials. However, it was not until astronauts on Gemini flights in 1966 experienced difficulties working outside their spacecraft that NASA took the problem seriously and resumed tests in the McDonogh pool to find ways of restraining an astronaut at a work station to prevent over-exertion and physical exhaustion.

Recognising that simulating planned spacewalks and rehearsing how best to carry out complex tasks was wise, astronauts began using the Water Immersion Facility in Building 5 at JSC. A pool 25ft (7.62m) in diameter and 16ft (4.9m) deep was used during the later Gemini and Apollo missions. The Neutral Buoyancy Simulator at NASA's Marshall Space Flight Center (which see) closed in 1979 and in preparation for the Shuttle programme JSC got its own Weightless Environment Training Facility

(WETF). Situated in Building 29, the WETF was 78ft (24m) by 33ft (10m) by 25ft (7.6m) deep and served until 1998.

NASA planned a bespoke facility which would have had a length of 235ft (72m), a width of 135ft (41m) and a depth of 60ft (18m) but the costs were too high. Nevertheless, in need of a facility capable of taking full size mock-ups of parts of the ISS, the agency purchased an existing facility from McDonnell Douglas and converted it into the Neutral Buoyancy Laboratory (NBL) located at the Sonny Carter Training Facility. Named after astronaut Sonny Carter who flew on STS-33 in 1989 but lost his life in a plane crash in 1991, the NBL is 202ft (62m) in length, 102ft (31m) wide and 40.5ft (12.34m) deep.

It has been used consistently since its official opening in April 1995, with mock-ups of ISS modules, truss assemblies, vehicles from the European Space Agency, Japan's space agency and SpaceX with its Dragon capsules as well as the Cygnus cargo module from Northrop Grumman Innovation Systems (formerly Orbital ATK). There are drawbacks with using neutral buoyancy to simulate weightlessness, induced drag from water being one, which can lead to some disorientation and distraction and also from discomfort but generally it provides a better fidelity than trying to simulate in a dry environment.

Simulating the space environment is essential for qualifying spacecraft for operations in space and the Space Environment Simulation Laboratory achieves that from Building 32. It consists of two chambers, Chamber A with a diameter of 45ft (16.7m), a height of 90ft (27.7m) and a circular floor which can rotate through 180°, four overhead cranes each capable of lifting 50,000lb (22,680kg), with primary equipment lifted in or out of the chamber by a 100,000lb (45,360kg) external crane. Chamber A has a solar lighting array and the ability to provide a thermal plasma field which effectively simulates the space environment outside the Earth's atmosphere. Two airlocks allow human access, one at ground level and a second at a height of 31ft (9.4m), which double for access to human occupants on test runs as well as altitude tests facilitating reduced atmospheric pressure.

Chamber B is smaller, with a diameter of 25ft (7.6m) and a height of 26ft (7.9m), which

has use of two 100,000lb (45,360kg) cranes and it too has two airlocks, one of which has a water deluge capability for simulating oxygen rich environments but the solar lighting simulation is a lot less sophisticated than Chamber A, a mirror system being used to achieve required lighting angles for simulation effects. The advantage of Chamber B is that it is more effective when testing smaller objects or structures and the simplicity of operation allows quicker turnaround. Both Chamber A and B can express a temperature of -300°F (-184°C) to an upper level depending on the specific test article and a pressure range of 1 x 10-6 to 760 torr (1 torr is one standard atmospheric pressure).

For integrated environment testing, Chamber E has a diameter of 4.6ft (1.4m), a height of 9.5ft (2.9m) and creates a thermal-vacuum environment for large gas loads at high vacuum, capable of dropping the temperature down to -280°F (-173°C) and a pressure range between 1 x 10-6 and 760 torr. Several other simulation chambers are available for measuring thermal bakeout and several levels of vacuum testing, all of which can contribute toward space suit development, life-support equipment testing and the separate elements in an environmental control system for manned space vehicles.

Integrated environment testing allows a wide range of simulations to be applied to spacecraft which are designed for operating on planetary surfaces, providing evaluation of seals, gaps, attachment devices and other items of hardware to test how they react in the environment for which they were designed. These tests range all the way from simulated ascent heating through the atmosphere to cold soak in space for extended periods. Aerothermal heating as spacecraft enter planetary atmospheres is tested using the 13MW arc tunnel which allows wide variations in the simulated environment to be experienced in this high-altitude, hypersonic wind tunnel.

A significant amount of work is also carried out on robotics and the development of associated hardware using a motion platform, the origin of which dates back to the days of the Gemini programme when an air platform allowed freedom in three degrees of motion for determining the reactive force upon an

astronaut using a hand-held manoeuvring unit, developed very early in NASA's experience with EVA in the belief that a reactive gas-gun could be used to translate from place to place outside a space vehicle. The equipment is different and the technology a lot more sophisticated but the motion-base simulator now available for users can manoeuvre loads of up to 500lb (226.8kg) at the end of a 60ft (18.2m) articulated arm.

The dextrous manipulator used on the ISS has a test bed down on Earth in the form of two 6-point hydraulic manipulators with control systems attached to a 7ft (2.1m) high pedestal. It can be used to guide with precision loads of up to 240lb (108.8kg) for manoeuvring in small spaces. Robotic arms and manipulators are destined to play an increasingly important role as NASA supports future space projects where humans are integrated fully with robots and autonomous machines working both inside and outside space vehicles.

The Johnson Space Center employs around 3,200 personnel and actively engages with several other NASA field centres in its primary goal of managing manned space flight development and operations. It projects expertise in human space vehicle systems, environmental control and life support systems, in EVA and related technologies and in flight design and planning. Supporting all these activities are the integrated environmental testing facilities outlined above.

Public access to the Johnson Space Center is via the Space Center Houston, a non-profit organisation which opened in 1992

BELOW A panoramic view of the Space Vehicle Mockup Facility which changes significantly over time to provide full-scale training in a dry, 1g environment for technicians, crew and flight managers. *(NASA)*

and contains a wide variety of exhibits and themed displays as well as a movie theatre screening a selection of space films showing the achievements of the NASA facility. An open-air tram tour takes in Mission Control, a restored Saturn V lying on its side and to a variety of locations, access to which is dependent on centre activity. But a special Level 9 tour limited to 12 people at a time lasts more than four hours and provides access to a wide range of facilities, to Mission Control and to some of the engineering test areas which would not normally be accessible to the general public.

Lunar Sample Laboratory Facility

Preparations for receiving back on Earth the first lunar samples began in the early 1960s and by 1964 the first plans were formulated for what would emerge as the Lunar Receiving Laboratory (LRL), located at JSC as Building 37. This 86,000ft^2 (8,000m^2) facility was completed in 1967 and was used for samples from the Moon where preliminary analysis could be carried out to characterise the materials. It also disseminated samples to teams of research workers around the world, based on the integrity of their research. However, it lacked storage and a new Sample Storage and Processing Laboratory (SSPL) was set up in Building 31, to which all Apollo mission samples were transferred after the end of the programme.

Concerns about the vulnerability of this priceless collection in a single facility led to a distribution among several separate buildings at JSC and at a secondary site located at Brooks Air Force Base in San Antonio, Texas. When that facility closed in 2002 the small secondary collection was moved from there to White Sands Test Facility, where the 115lb (52kg) of lunar material is now contained; the remaining 727lb (330kg) of primary samples were housed in a completely new facility at JSC known as the Lunar Sample Laboratory Facility (LSLF).

Constructed as an annex to Building 31 and known as 31N, the LSLF was dedicated on 20 July 1979, exactly a decade after the first manned lunar landing. It consists of a two-storey building with 14,000ft^2 (1,300m^2) of floor space, storage vaults, laboratories for

preparing and analysing samples and special cabinets fed with nitrogen for long-term storage. The entire building was constructed with a view to eliminating contamination, air filtered in to remove suspended particles and a slight increase in atmospheric pressure to prevent air from the outside getting in. Access is only permitted in cleanroom suits and samples are handled only through multi-layered gloves. When samples have been sent out and worked on they are returned to a "non-pristine" section of the facility to prevent contamination. Storage capacity greatly exceeds present demand, in anticipation of a resumption of manned expeditions to the Moon.

The LSLF contains some of the most remarkable rocks, lunar soil and materials from the surface of the Moon, among samples from six Apollo landings sites harvesting 842lb (382kg) consisting of 2,200 separate, catalogued pieces, of which 75% are contained at the LSLF in pristine condition. The rest have been processed into smaller pieces comprising 110,000 samples individually catalogued. Two of arguably the most priceless geological specimens on Earth are the Genesis Rock, with an age of 4.1billion years and Big Muley, named after Apollo 16 geology team leader Bill Muelberger which, at 25.9lb (11.745kg), is the largest rock brought back from the Moon.

ABOVE The prize of the century! The first box of lunar samples retrieved from the Moon by the crew of Apollo 11 arrives at Building 37, then known as the Lunar Receiving Laboratory. *(NASA)*

3 Vandenberg Air Force base

While Cape Canaveral continues to dominate the news about space launches from the East Coast of the United States, Vandenberg Air Force Base (VAFB) on the West Coast has played at least as important a role in the development and expansion of US missile and space activity since the first rockets were fired from there more than 60 years ago. Unlike Cape Canaveral, however, VAFB is not a publicly accessible space and has therefore not received the same attention. As the site for some important NASA launches, it is nevertheless a vital part of NASA operations.

OPPOSITE At SLC-4E, a Falcon 9 rocket stands ready to launch Iridium communication satellites into orbit. *(SpaceX)*

Located 163 miles 262km) up the coast north of Los Angeles, California, the sprawling location of the current VAFB was procured for the Army in March 1941 and named Camp Cooke, in honour of Maj Gen Philip St George Cooke, a Union soldier in the American Civil War. Construction began in September that year with activation on 5 October, following which it was used for a variety of training activities until closed in June 1946 only to be reopened in August 1950 for the duration of the Korean War and closed again in February 1953.

With the advent of rocketry and missile development, Camp Cooke was reactivated in November 1956 when the Army transferred 64,000 acres (25,900ha) to the Air Force for launch facilities. The first launch took place when a Thor IRBM was fired from pad 75-1-1 on 16 December 1956. It was known as Cooke Air Force Base and in October 1957 was assigned to Strategic Air Command for testing ballistic missiles, receiving its present name on 4 October 1958 in honour of Gen

Hoyt S. Vandenberg, the second Air Force Chief of Staff.

The southern portion of the facility had been transferred to the Navy in May 1958 for a missile launch facility at Point Arguello, transferred back to the Air Force in July 1964 with the Air Force acquiring various other outlying border regions of the site in February 1965. Further land acquisition occurred the following March when Space Launch Complex 6 (SLC-6) was planned as the launch site for the ambitious Manned Orbiting Laboratory, a classified military space station formally approved in 1965 but cancelled in June 1969, after which it was later converted into a launch facility for the Shuttle. The Air Force never used it for that reusable vehicle and today it has been leased by Boeing for flights with the Delta IV rocket on polar-orbit military missions (see SLC-6 on page 204).

The first space launch from VAFB began a wide range and continuing series of classified military satellite flights beginning on 28 February 1959 with the first Corona

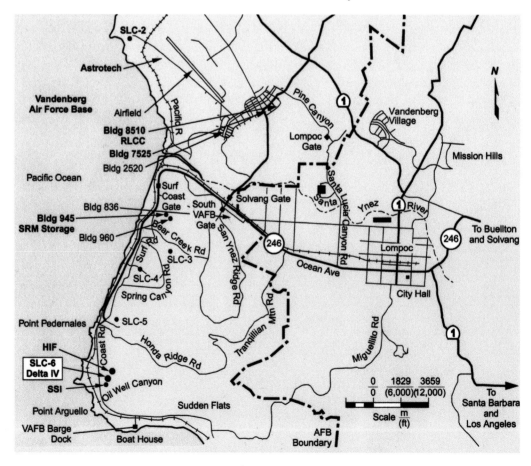

RIGHT A map of Vandenberg Air Force Base where most flights are for military satellites with NASA using some pads for polar-orbit flights. *(ULA)*

spy satellite. VAFB is uniquely positioned to launch satellites into Sun-synchronous orbit where the satellite can be delivered to a near-polar orbit so that the Sun falls across the same area of the Earth's surface observed by the satellite every day, illuminating the same places. This is valuable for a wide variety of applications including spy satellites observing the same place over a protracted period of time, weather observation so that it can provide continuous observations of the same strip of latitude every day, or for observing the Earth's natural resources.

Because launch vehicles discard spent boosters or rocket stages, they are not allowed to fly over land in the early phases of their trajectory toward orbit. For this reason flights from Cape Canaveral are limited to a flight azimuth which does not cross land areas of the Eastern seaboard. Flight to high-inclination orbits, especially to polar orbits, are required to make a plane change as they leave the atmosphere and "turn" their trajectory to a 90° orbital inclination. That is expensive on energy in the launch vehicle and dramatically reduces the weight of the payload to compensate.

Because of this, NASA began using VAFB for its polar-orbit satellites, with flights from the West Coast site launching due south, over the South Pole, back up to the North Pole and around the Earth, past the latitude of VAFB and so on for a second orbit. However, because the Earth rotates in a counter-clockwise direction as viewed from the North Pole, the ground-track of a polar-orbiting satellite shifts approximately 1,556 miles (2,500km) west for each 90-minute orbit. It would only pass over VAFB again 24 hours after launch.

Due to the nature of VAFB, and the classified mission of many of the orbital launches taking place across the facility, when NASA began launching polar orbiting satellites from there the intention was from the outset to contain non-military satellite flights from a few designated launch pads. Non-government personnel were heavily involved both in the satellite and payloads and the preparation of hardware for launch so access could be contained to those areas cleared for unclassified flights.

Vandenberg Air Force Base launch pads

SLC-2E (16 December 1958) 34.7516°N x 120.6193°W

This complex was built in 1957 to support development and test flights of the Thor IRBM, both for the US Air Force and for the UK's Royal Air Force, which launched three Thor missiles from this pad between 3 August and 2 December 1959. In 1960 it was modified to accept the Thor-Agena B with the first launch of that vehicle on 16 June 1961. From 28 June 1962 it supported launches of the Thor-Agena D and from 28 September 1963, Thor-Able Star launchers.

Until the first NASA launch from VAFB on 29 September 1962 when it managed the flight of Canada's first home-built satellite, Alouette 1 for research into the ionosphere on a Thor-Agena B, all launches from SLC-2E had been for the military but the increasing interest in polar-orbiting satellites equipped for observing the Earth and near-Earth space from this orbit opened the site for increasing NASA participation over the years to come.

The next non-military launch from SLC-2E was the flight of Nimbus A, a weather satellite developed at the Goddard Space Flight Center, sent up to near-polar orbit on 28 August 1964. This was followed by the second Orbiting Geophysical Observatory (OGO-2) on 14 October 1965. There followed a series of different applications satellites for NASA until the final Thor-derivative flight from this complex on 12 March 1972, the TD-1A navigation technology research satellite for the European Space Research Organisation (ESRO), precursor to the European Space Agency (ESA). LC-2E was abandoned in 1975 and the site stripped of its gantry.

SLC-2W (17 September 1959) 34.7556°N x 120.6223°W

Built as a Thor IRBM test facility for the US Air Force, it was also used to train the UK's RAF Bomber Command crews who would man the missile in Britain, the first three Thor launches being with RAF crews. The first orbital flight occurred on 29 August 1962 with the launch of a Corona spy satellite of which several more of type were launched.

The first NASA mission was on 23 January 1970 when ITOS-1, a weather satellite developed from the Tiros series, was launched from this site. Notable too was the launch of NASA's first dedicated Earth resources satellite, Landsat 1, launched from here on 15 October 1972.

Landsat 2 followed on 22 January 1975, Landsat 3 on 5 March 1978, Landsat 4 on 16 July 1982 and Landsat 5 on 1 March 1984, interspersed with several scientific, environmental, resource monitoring and meteorological satellites. SLC-2W saw the growth and expansion of the Delta programme, based on the Thor IRBM, and evolved to support an expanding use of commercial companies operating telecommunication satellite services, including Iridium which began launching clusters of these low-orbit payloads in fives on 5 May 1997.

Successive launches of Iridium clusters flew in groups of five over the next ten flights until the last from this complex on 6 November 1998 by which time 55 had been placed in orbit. A further cluster of five went up on 11 February 2002. A wide selection of scientific and applications satellites continued to fly until the Delta launch vehicle evolved to a very different configuration which required a bigger launch complex, the Delta IV which first flew from Cape Canaveral's LC-37 (which see) on 20 November 2002 and from SLC-6 at Vandenberg on 27 June 2006. The last launch of a Delta II configuration from SLC-2W occurred on 18 November 2017.

SLC-3W (11 October 1960)
34.6400°N x 120.5911°W

Built initially to support Atlas-Agena A (of which only two were flown) and B series satellite launchers (of which seven were flown), this was the first dedicated satellite launch facility at VAFB. Initially designated LC-1, modifications to the site took place in 1963 with a complete rebuild to allow for the launch of more advanced Thor-Agena rockets and it was refurbished again in 1973 for the launch of Atlas variants with upper stages.

Until the launch of the Seasat 1 satellite on 27 June 1978, SLC-2W was used exclusively for the launch of military payloads, principally the Corona series and later types of intelligence gathering and navigation satellites. Carrying a set of microwave instruments for measuring ocean currents, sea heights and oceanographic conditions, Seasat 1 was launched by Atlas-Agena D but operated only 99 days before an electrical fault rendered it inoperative.

After Seasat there followed a wide range of environmental, meteorological, scientific and Earth resource satellites, interspersed with military launches to near-polar orbits. The last of 81 Atlas and Thor launches from this facility took place on 24 March 1995.

In 2004, the commercial launch provider SpaceX began modifying the site for the Falcon 1 rocket which the company would use to static fire the first stage but only one test took place, on 3 May 2005, before trials moved to Omelek Island.

SLC-3E (12 July 1961)
34.6359°N x 120.5878°W

Built to support flights for space launches, first with Atlas-Agena B, then Atlas-Agena D and Atlas-Burner II flights, the launch facility also supported Air Force Prime launches carrying a semi-lifting re-entry body on aerothermal tests during re-entry. However, the first launch from this site placed a Midas defence early warning satellite in orbit, followed by photographic reconnaissance satellites on the Samos series, scientific research satellites and the first seven Navstar GPS satellites in sequence between 22 February 1978 and 19 December 1981.

NASA's first launch from SLC-3E was the Terra satellite on 18 December 1999 carried on the first Atlas IIAS launcher. Seven years later modifications began for launching the Atlas V, the first launch of which occurred on 13 March 2003. The Atlas V programme was part of the EELV programme (see SLC-6 on page 204) and this was the first flight of type.

On 13 August 2014 an Atlas V launched the commercial imaging satellite Worldview-3 with a ground resolution of 1.2ft (36.6cm) in addition to multispectral and infrared imaging equipment. Worldview-4 followed on 11 November 2016 carried by another Atlas V and accompanied by seven Cubesats. In a historic launch, NASA's first interplanetary probe launched from the

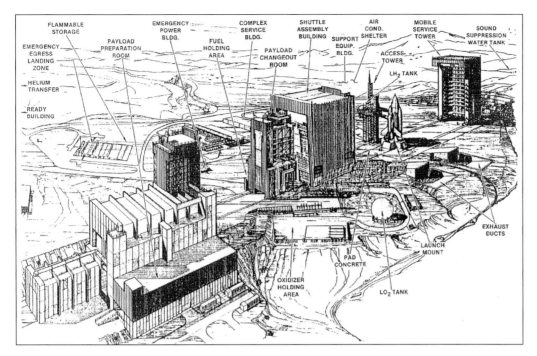

Labels on illustration: FLAMMABLE STORAGE, EMERGENCY EGRESS LANDING ZONE, HELIUM TRANSFER, READY BUILDING, PAYLOAD PREPARATION ROOM, EMERGENCY POWER BLDG., FUEL HOLDING AREA, COMPLEX SERVICE BLDG., PAYLOAD CHANGEOUT ROOM, SHUTTLE ASSEMBLY BUILDING, SUPPORT EQUIP. BLDG., AIR COND. SHELTER, ACCESS TOWER, LH₂ TANK, MOBILE SERVICE TOWER, SOUND SUPPRESSION WATER TANK, EXHAUST DUCTS, LAUNCH MOUNT, PAD CONCRETE, OXIDIZER HOLDING AREA, LO₂ TANK

West Coast was sent on its way to Mars on 5 May 2018 when the InSIGHT began a six-month flight and a landing scheduled for late November the same year, when it is hoped it will become the eighth NASA spacecraft to land successfully on the Red Planet.

SLC-5 (26 April 1962)
34.6080°N x 120.6247°W

Built specifically for the all-solid Scout small satellite launcher, the first launch carried the US Navy's solar radiation satellite Solrad but the attempt was a failure, the first successful launch occurring on 23 August 1962 after a second failure a month earlier. In total some 69 Scout rockets were launched from SLC-5, the last on 8 May 1994 after which the pad was deactivated.

The first NASA launch was on 19 December 1963, a landmark event since this was the first time the agency had used VAFB to send up one of its own satellites. Only Canada's Alouette 1 satellite launched from SLC-2E on 29 September 1962 preceded it but that had not been an in-house NASA programme. Explorer 19 was designed to measure the density of the ionosphere and was followed on 25 August by Explorer 20.

Over the next several years SLC-5 saw the launch of more Explorer-series NASA science satellites and some for foreign countries, ESRO-1B being launched for the European Space Research Organisation on 8 November 1969, while several military navigation satellites were sent into orbit.

SLC-576E (1 August 1962)
34.7396°N x 120.6191°W

A most unusual launch site in that it consisted originally of a single silo for the Atlas F ICBM, a variant inaugurated here and from which four test launches were made, the last on 22 December 1964. For the next 30 years the site remained dormant, decommissioned and stripped of equipment until it was modified for flights with the commercial Taurus launch vehicle.

Developed by Orbital Sciences and now a part of Northrop Innovation Systems, Taurus was a four-stage ground launcher developed from the air-launched Pegasus, a rocket dropped from the underside of a converted Lockheed L-1011 airliner. The first Taurus launch occurred on 13 March 1994, the last of ten on 31 October 2017 by which time Taurus had been renamed Minotaur. A total of 11 flights had been completed by 31 October 2017.

SLC-4W (12 July 1963)
34.6331°N x 120.6154°W

Built specifically for the Atlas-Agena D in 1962, this launch complex saw a consistent series of flights for classified military payloads. After the last launch of 12 Atlas-Agena D from this complex on 12 March 1965, modifications were made for it to convert to classified payloads launched on derivative launcher variants of the Titan II ICBM, going by a variety of designations according to the specific configuration and upper stage. The first Titan launch took place on 29 July 1966.

The first non-military launch occurred when NASA flew its Landsat 6 remote sensing satellite into near-polar orbit on 5 October 1993. However, the kick-motor designed to place it in its final orbit failed. Next up on 25 January 1994 was the Clementine mission which had a primary function of testing sensor technology for the so-called Star Wars ballistic missile defence system but a highly valued secondary role of orbiting the Moon and carrying out a mapping survey of the lunar surface.

A mix of some defence and weather satellites followed, with NASA flying QuickScat on 20 June 1999, an oceanographic satellite incorporating a SeaWinds scatterometer for measuring and monitoring winds across the world's oceans. The last launch occurred on 18 October 2003, a defence weather satellite, after which the complex was deactivated.

SLC-4E (14 August 1964) 34.6320°N x 120.6106°W

Built in 1963 for the Atlas-Agena D satellite launcher, a total of 27 were launched by 4 June 1967 after which modifications were made for launches with the Titan IIID, of which 22 were flown between 15 June 1971 and 17 November 1982. Further upgrades were made to allow it to support the Titan 34D, seven of type being launched between 20 June 1983 and 6 November 1988. Changes adapted the site to support eight Titan IVA launches between 8 March 1991 and 24 October 1997 after which it was modified for the Titan IVB supporting five launches between 22 May 1999 and 19 October 2005.

In early 2011, SpaceX began rebuilding the site to support their Falcon 9 and Falcon Heavy launchers, although the latter would never fly from there. Modifications included removing the very large fixed and mobile service towers which had been gradually built up over the previous decades to support increasingly large Titan derivatives. The first launch of a Falcon 9 v1.1

took place on 29 September 2013 carrying the Canadian communications satellite Cassiope and several Cubesats. Next up was the launch of Jason-3 on 17 January 2016, a cooperative programme between NASA and the European weather satellite organisation Eumetsat. On 14 January 2017 a Falcon 9 launched the first cluster of second-generation Iridium satellites. SLC-4E continues in use for Falcon 9 launches with a manifest currently booked through 2021.

SLC-6 (15 August 1995) 34.5813°N x 120.6266°W

This launch complex has a very strong association with NASA despite the fact that no agency mission ever launched from here. It was planned as the launch site for the US Air Force Manned Orbiting Laboratory, a military space station which was to comprise a cylindrical structure with a two-man crew launched in a Gemini spacecraft on top of the MOL itself.

Initially, the launch vehicle was to have been the Saturn I but a change to a developed version of the Titan II with strap-on boosters was made for which land was prepared for SLC-6 beginning on 12 March 1966. The Titan IIIM launch site would have a Mobile Service Tower, a concrete launch pad, a flame duct and a Launch Control Center. On 10 June 1969 the MOL was cancelled and work on the site stopped. When NASA got approval to build the Shuttle in 1972, SLC-6 was selected as the future launch site for polar missions for that reusable vehicle, allowing it to fly high-inclination orbits without overflying populated areas.

The Air Force had been closely involved in the design configuration of the Shuttle and wanted to deliver to orbit highly advanced spy satellites completely different to anything which had flown before and that required a big launch system and a new capability on the West Coast. At one time it was mooted that the Shuttle could rendezvous with previously launched spy satellites and replenish their propellant tanks, extending the life of these expensive platforms.

Work began in January 1979, later than desired due to NASA's budget pressures, on a much more comprehensive facility than had ever been envisaged for the MOL. Shuttle Orbiters would be delivered to

North Vandenberg by air atop a converted Boeing 747 and moved south to SLC-6 on a 76-wheel flatbed truck where it would be erected to a vertical position for launch. All the payload changeout facilities created for the Shuttle at Cape Canaveral would be duplicated here but under a very different operational procedure.

NASA and the Air Force spent more than $4billion on rebuilding SLC-6 for Shuttle operations and the Orbiter *Enterprise*, which previously had conducted Air Launched Tests to demonstrate the integrity of the vehicle, was taken to SLC-6 in 1985 for a fit-check at the facility. After the *Challenger* disaster of 28 January 1986, the Air Force decided not to use the Shuttle for classified missions from VAFB and on 31 July it announced that it would close the SLC-6, reducing it to minimum caretaker status on 20 February 1987.

It had been a close run thing; had *Challenger* not been destroyed the first Shuttle launch from VAFB would have taken place on 15 October 1986 on a mission commanded by Robert "Bob" Crippen, one of the astronauts originally selected for the MOL programme. The Air Force had never placed great faith in the Shuttle and after the accident it reverted to the Titan launch vehicle and made plans to launch the Titan IV/Centaur from SLC-6. But the Air Force cancelled those plans on 22 March 1991 and the facility was once again abandoned.

In 1993, however, Lockheed Martin reached an agreement to fly its Lockheed Martin Launch Vehicle (LMLV) from SLC-6 using a "milk-stool" pedestal over one of the two exhaust ducts built for the Shuttle solid rocket boosters. The first LMLV took off on 15 August 1995, the first rocket launch from this site for more than 29 years after it had been selected for the MOL programme. Unfortunately, the flight was a failure. However, on 22 August 1997 NASA's Small Satellite Technology Initiative (SSTI) satellite, a part of its "Mission to Planet Earth" initiative was successfully placed in orbit by a variant of the LMLV called Athena, for which some pad modifications had been necessary. On 24 September 1999 the Ikonos imaging satellite was launched, three months after a precursor of the same type failed to make it into orbit.

As noted elsewhere, Boeing began a move to retire the now ageing Delta II and introduce the Delta IV series of several adaptable core stages, strap-on boosters and a variety of upper stages. This was part of a competition run by the Defense Department to find a common solution to military satellite launcher requirements, identified as the Evolved Expendable Launch Vehicle (EELV) programme. It was all a part of the dramatic changes following the Shuttle accident, when all but remaining manifested launches such as planetary flights and the Hubble Space Telescope, were cancelled to focus the Shuttle on preparing for the assembly of the International Space Station.

The first Delta IV to fly from VAFB was taken to SLC-6 and erected on the pad in late 2003 but successive modifications and technical challenges delayed the launch until 27 June 2006, the sixth Delta IV to fly, the first having been from SLC-37B at Cape Canaveral on 20 November 2002. Flights continue from SLC-6 supporting a wide range of classified military satellite programmes.

BELOW The first Delta IV-Heavy lifts off from SLC-6 on 28 August 2013 in its present role as launch facility for United Launch Alliance. *(USAF)*

4 The NASA budget

When asked how much the space programme costs, most people get it wrong. The average answer is around 20% of total government spending each year – in other words, around 20 cents in every tax dollar. In fact, NASA takes less than 0.5% out of the federal coffers each year and plies back around ten times that into the economy through a variety of technology feedback activity which enhances commercial and corporate jobs, profits and export potential. Independent analysis shows that this sum boosts the economy by approximately $180–200billion each year.

OPPOSITE NASA is planning to send astronauts deeper into space than at any time since the last Apollo Moon flight in 1972. That costs money. But just how much? *(NASA)*

Taken as an example, in 2019 NASA will receive around $20billion, or 0.4% of the $4,407billion federal budget – not GDP, just the amount the US government spends. By comparison, the Department of Defense and Homeland Security together will spend $643billion, almost 15% of the budget. In that comparison, the same amount of money would fund NASA at its present level for more than 32 years.

In another example, in 2019 the US government will spend $158.6billion on health and human services, education and housing and urban development, 3.6% of its total budget. If NASA's budget were stopped and applied to these social welfare programmes, it would increase funds for those worthwhile sectors by 11% but deprive the US economy of $160–180 billion a year (a figure derived from the $180–200 billion value to the economy minus the $20 billion spent on NASA to get that benefit).

By every conceivable measure, NASA alone more than funds the entire budget for health and human services, education and housing and urban development.

But it has not always been like that. From very low beginnings, NASA's budget grew almost exponentially, reaching 4.4% of total federal expenditure by 1966, this being due to the enormous investment made in the Apollo Moon landing goal and the hardware, buildings and facilities construction essential to support that. In several respects, the benefits passed on to successive programmes was instrumental in helping NASA achieve remarkable results with a declining budget, which it did from 1966, falling to less than 1% of federal expenditure each year by 1975, slowly declining to less than 0.5% since then.

While there has generally been universal support in both Houses of Congress for what it is doing, NASA loses out when it comes to its budget allocation through the lack of real perception in the public awareness of the extraordinary benefits that accrue to the nation from its investment. This feeds through to local political interests when it comes to satisfying the electorate. That, and the increasing wealth of the United States, plus the readiness to run a budget deficit to a far greater level than existed until the last 20 years, has prevented a real-term rise and contributed toward a decline when measured as a percentage of annual government expenditure.

Deciphering government expenditure is a tough call, full of statistical minefields and conduits that frequently lead to cul-de-sacs! But essentially, a major challenge for Congress

BELOW The NASA budget has taken a consistently falling percentage of the US federal spending plan each year while the return to the economy pays for all annual expenditure on health, human services, education and housing and urban development.
(Via David Baker)

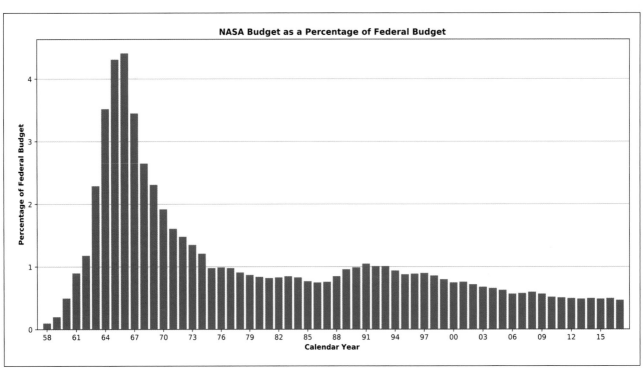

when it comes to finding more money is that the budget is really divided into two sectors: non-discretionary and discretionary spending. Non-discretionary spending categories contain all the programmes that are mandated by Congress, many with built-in escalators, which automatically eat up increasing proportions of the budget each year. Discretionary spending is the amount remaining for distribution each year for which there is no legally binding level.

Discretionary spending has plummeted in the last 60 years of NASA budgets, squeezing the available remaining after the untouchable non-discretionary payments have been made. The pressure on the defence budget is an example. During the Second World War the US spent 95% of its income on the conflict. Even during the Korean War of 1950–53, total defence spending was almost 70% of the total government budget. Now, it is down to a mere 13%.

For 2019, the government requested a discretionary budget of $1,200 billion, a mere 27% of total budget outlay. But even measuring NASA's budget by that yardstick, it gets only 1.6% of discretionary spending. With the returns it accrues to the economy, financial analysts have to work out whether more spent on NASA would bring financial returns proportionate to the increased investment in the nation's space programmes, or whether it has plateaued and would return less as a multiple of the agency's annual budget.

In realistic terms, it is highly unlikely that NASA will get more than around 0.5% of federal spending each year, despite being asked to do more by politicians not only seeing in NASA a highly beneficial activity for the nation, but also as a means of bringing jobs and employment to their constituency. Which throws a lot on to public awareness and the need for NASA to satisfy the preferences of the electorate.

Toward that end, a major national opinion poll conducted in early 2018 came up with some interesting statistics. About 70% think that NASA should strive to remain a world leader in space exploration, with 65% believing that the agency should retain a key role in exploration and scientific discovery, choosing to be modestly distrustful of the commercial contenders. Perhaps surprisingly, 85% of those asked felt that monitoring climate was a priority, followed by monitoring asteroids for their potential for inflicting catastrophic effects on impact.

When it came to human space flight, however, most remained unconvinced. Some 63% felt that sending astronauts to Mars was important and should be a priority but only 55% thought that human flights to the Moon were a priority. In fact 37% said that flights to Mars should not be on the agenda while 44% were set against sending astronauts to the Moon. Whether those levels of preference will change when NASA embarks on its first deep-space expeditions since Apollo remains to be seen. But overall, taxpayers in the United States believe in NASA and in its vital role extending our understanding of the planetary environment.

Index